OPTICAL FIBERS: MATERIALS AND FABRICATION

Advances in Optoelectronics (ADOP)

Editor: T. OKOSHI (*Univ. of Tokyo*)

Associate Editor: T. KAMIYA (*Univ. of Tokyo*)

Editorial Board:

G. A. ACKET (*Philips Res. Labs., The Netherlands*)
J. A. ARNAUD (*Univ. Limoges, France*)
S. A. BENTON (*Mass. Inst. Technol., U.S.A.*)
W. A. GAMBLING (*Univ. of Southampton, England*)
T. G. GIALLORENZI (*Naval Res. Lab., U.S.A.*)
J. W. GOODMAN (*Stanford Univ., U.S.A.*)
I. HAYASHI (*Optoelectr. Joint Res. Lab., Japan*)
H. INABA (*Tohoku Univ., Japan*)
E. A. J. MARCATILI (*AT&T Bell Labs., U.S.A.*)
Y. SUEMATSU (*Tokyo Inst. Technol., Japan*)
J. TSUJIUCHI (*Tokyo Inst. Technol., Japan*)
R. ULRICH (*Tech. Univ. Hamburg-Harburg, West Germany*)
H.-G. UNGER (*Tech. Univ. Braunschweig, West Germany*)
Emil WOLF (*Univ. of Rochester, U.S.A.*)
H. YANAI (*Toshiba Corp. & Shibaura Inst. Technol., Japan*)

ADOP Advances in Optoelectronics

OPTICAL FIBERS: MATERIALS AND FABRICATION

T. Izawa and S. Sudo

Nippon Telegraph and Telephone Corporation, Japan

KTK Scientific Publishers/Tokyo

D. Reidel Publishing Company

A MEMBER OF THE KLUWER ACADEMIC PUBLISHERS GROUP

Dordrecht / Boston / Lancaster / Tokyo

Library of Congress Cataloging-in-Publication Data

Izawa, Tatsuo.
   Optical fibers.

   (Advances in optoelectronics)
   1. Optical fibers.   2. Fiber optics.   I. Sudo,
Shoichi.   II. Title.   III. Series.
TA1800.I93   1986   621.36′92   86-26075
ISBN 90-277-2378-8

Published by KTK Scientific Publishers (KTK),
307 Shibuyadai-haim, 4-17 Sakuragaoka-cho, Shibuya-ku, Tokyo 150, Japan,
in co-publication with D. Reidel Publishing Company, Dordrecht, Holland

Sold and distributed in the U.S.A. and Canada
by Kluwer Academic Publishers,
101 Philip Drive, Assinippi Park, Norwell, MA 02061, U.S.A.
in Japan by KTK Scientific Publishers (KTK),
307 Shibuyadai-haim, 4-17 Sakuragaoka-cho, Shibuya-ku, Tokyo 150, Japan

In all other countries, sold and distributed
by Kluwer Academic Publishers Group,
P.O. Box 322, 3300 AH Dordrecht, Holland

All Rights Reserved

Copyright © 1987 by KTK Scientific Publishers (KTK)

No part of the material protected by this copyright notice may be reproduced or utilized in any form or by any means, electronic or mechanical, including photocopying, recording or by any informational storage and retrieval system, without written permission from the copyright owner.

Printed in Japan

PREFACE

Many scientists have been attracted to the interaction phenomena between photon and solid-state materials, however, very few scientists have studied the interaction minutely. Therefore, it is rather difficult to answer simple questions such as what is the most transparent solid-state material and what is the limit of its transparency.

Ever since the prediction of the possibility of low-loss optical fibers by C. K. Kao, scientists and electronic engineers have made great efforts towards its realization. Most of them have concentrated on the selection of fiber materials and the development of fabrication processes. Doped silica glass fibers fabricated by the OVD, MCVD or VAD processes have survived worldwide competition. Today, we can obtain silica glass fiber with a transmission loss as low as 0.15 dB/km. There is still however the dream of an ultralow-loss fiber which would make it possible to communicate across the Pacific Ocean or the Atlantic Ocean without the use of a repeater.

Many kinds of textbooks and monographs on optical fibers have been published throughout the development of low-loss optical fibers, most of them have been devoted only to the theoretical aspects of optical waveguides. This book describes the practical aspects of optical fibers such as fiber materials and fabrication processes. Detailed data on the optical, thermal, chemical and mechanical properties of various fiber materials measured by the authors and their colleagues at NTT Ibaraki Laboratories are described. The fabrication processes of optical fibers are also described from the viewpoint of understanding what factors are important in the manufacture of low-loss fibers. The VAD process, which has been developed by the authors, is described in detail. This description will be valuable not only for researchers but also for communication system designers, cable designers and other related engineers.

Although the authors are pessimistic with regard to the realization of an ultralow-loss fiber of less than 0.01 dB/km, we believe that this book will assist in clarifying both the important and the significant factors in the manufacture of low-loss fibers.

The authors wish to thank their colleagues at NTT Ibaraki Laboratories for their provision of various data on fiber materials and fabrication processes. The authors are much indebted to Drs. N. Shibata, T. Miya, S. Sakaguchi and S. Takahashi for making theses available to us. The authors would like also to express their sincere gratitude to Professor T. Ohkoshi for his encouragement throughout the preparation of this work.

CONTENTS

Preface v

Chapter 1
OPTICAL-FIBER MATERIALS 1
1.1   Materials for optical fibers 1
1.1.1 Introduction 1
1.1.2 Oxide glass 4
1.1.3 Halide glass 6
1.1.4 Chalcogenide glass 6
1.1.5 Single crystals and polycrystals 6
1.1.6 Plastics 8
1.2   Optical transmission loss of fiber materials . . . 8
1.2.1 Introduction 8
1.2.2 Phonon absorption 11
1.2.3 Electron transition absorption 15
1.2.4 Weak absorption tail 16
1.2.5 Scattering loss 18
References 20

Chapter 2
OPTICAL PROPERTIES OF PURE AND DOPED SILICA
. . . . 23
2.1   Transmission loss 23
2.1.1 Introduction 23
2.1.2 Phonon absorption and multiphonon absorption . . . 23
A) SiO_2 23
B) GeO_2 26
C) P_2O_5 26
D) B_2O_3 27
2.1.3 Ultraviolet absorption and the tail 28
2.1.4 Impurity absorption 31
A)   Transition metal 31
B)   Hydroxyl ions 32
2.1.5 Scattering loss 33

2.1.6	Total loss of high silica fibers	35
2.2	Refractive-index	36
2.2.1	Refractive-index dispersion	37
2.2.2	Material dispersion	38
2.2.3	Profile dispersion	41
2.2.4	Temperature dependence	44
2.3	Linear thermal expansion coefficients	46
References		49

Chapter 3
FABRICATION PROCESS OF HIGH SILICA FIBERS 51

3.1	Introduction	51
3.2	MCVD process	54
3.2.1	Instruments for MCVD process	56
3.2.2	Tube diameter control	58
3.2.3	Prebake of silica tube	58
3.2.4	Hydroxyl impurity reduction	61
3.2.5	Optimum deposition temperature	63
3.3	Outside Vapor Deposition process	64
3.4	Fiber drawing	66
3.4.1	Drawing apparatus	67
3.4.2	Diameter control	68
3.4.3	Strength	69
A)	Surface treatment	69
B)	Dust particles in the furnace	70
C)	Humidity	71
D)	Primary coating	71
References		74

Chapter 4
VAPOR-PHASE AXIAL DEPOSITION PROCESS 77

4.1	Introduction	77
4.2	Preform fabrication apparatus	78
4.2.1	Porous preform fabrication chamber	80
A)	Torches	80
B)	Reaction chamber	81
4.2.2	Consolidation furnace	82
4.2.3	Pulling mechanism	83
4.2.4	Exhaust system	85
4.3	Porous preform fabrication	85
4.3.1	Synthesis and deposition of fine glass particles	86
4.3.2	Preparation and size control of porous preforms	88
4.3.3	Simultaneous cladding formation	90

4.3.4	High-speed production of porous preforms	94
4.4	Dehydration and consolidation	94
4.4.1	Experimental results	96
4.4.2	A model for final sintering stage	97
4.4.3	Consolidation Condition	101
A)	Gas permeability effect on pore behavior	101
B)	Temperature increasing speed	105
C)	Sintering in mixed gas atmosphere	106
4.5	Dehydration	107
4.5.1	Dehydration principle	107
4.5.2	Optimum dehydration temperature	109
4.5.3	Water vapor pressure	110
4.5.4	Dehydration reagent pressure	110
4.5.5	Dehydration time	111
4.5.6	Kinds of dehydration reagent	112
4.5.7	Cladding thickness	114
4.6	Profile control	115
4.6.1	Introduction	115
4.6.2	Dopant concentration	115
4.6.3	Profile formation mechanism	122
A)	SiO_2–TiO_2 system	122
B)	SiO_2–GeO_2 system	122
C)	Flame temperature effect	124
D)	Gas mixing effect	125
4.6.4	Profile control techniques	127
A)	Application of surface temperature effect	127
B)	Application of gas mixing effect	129
4.6.5	Fluorine doping	131
References		132

Chapter 5
FABRICATION PROCESSES OF MULTI-COMPONENT GLASS FIBERS AND NON-SILICA FIBERS 137

5.1	Multi-component glass fibers	137
5.1.1	Fabrication process	137
5.1.2	Purification of raw materials	138
5.1.3	Bubble formation	141
5.2	Fluoride glass fibers	143
5.3	Chalcogenide glass fibers	148
5.4	Crystalline fibers	150
5.5	Plastic fibers	151
References		153

Chapter 6
TRANSMISSION CHARACTERISTICS OF OPTICAL FIBERS
... 155

6.1	Single mode fibers	155
6.1.1	Transmission loss	155
6.1.2	Transmission bandwidth	156
A)	Characterization technique	156
B)	Dispersion of single mode fibers	156
C)	Dispersion-free single mode fibers in 1.5-μm region	157
D)	Doubly clad single-mode fibers	160
E)	High bit rate transmission experiments	161
6.1.3	Bending loss	162
A)	Dip effect	164
B)	Bending radius and refractive-index difference	165
C)	Theory	165
6.2	Graded index high-silica fiber	167
6.2.1	Transmission loss	167
6.2.2	Bandwidth	168
6.2.3	Refractive-index fluctuation	168
6.3	Al$_2$O$_3$-doped high-silica fibers	172
6.4	Loss increase by hydrogen permeation	174
6.5	Multi-component glass fibers	175
6.6	Non-silica fibers	177
6.6.1	Fluoride glass fibers	177
6.6.2	Chalcogenide glass fibers	178
6.6.3	Crystalline fibers	179
6.6.4	Plastic fibers	179
	References	183

Index ... 185

Chapter 1

OPTICAL-FIBER MATERIALS

Fundamental optical properties of solid state materials are described with a main emphasis on the choice of fiber materials. For the fabrication of practical optical fibers, there are many aspects to be considered, for example, optical loss, refractive-index, productibility of thin fiber, physical and chemical stabilities, fabrication cost and so on. The most important factor for the material choice is the optical loss properties of materials. Although the mechanism of optical loss at ultra-low loss region has not been fully understood, various experimental data in this chapter will assist the material choice for optical fiber use.

1.1   Materials for Optical Fibers

1.1.1   Introduction
    There are many aspects required for optical fiber materials: The major features to be considered are;
    1) Optical loss properties; intrinsic loss properties and the potentiality of extrinsic loss reduction are of vital importance.
    2) Refractive-index control; precise control of refractive-index profile in radial direction and the reduction of index fluctuation along the axial direction are necessary for fabricating high quality fibers.
    3) Shape control; Cross sectional shape and size, the surface finish and the fluctuation of the size along axial direction of fibers strongly affect the transmission characteristics of the fibers.
    All these three aspects are necessary for realizing high quality fibers. When the shape controllability of a material is poor, even if the intrinsic loss of a material is very small, it is hard to fabricate high quality fiber. Some kinds of crystalline material are in this category. Although we can find very low-loss materials in alkali-halide crystals, shape control techniques of crystalline materials has not been well developed.
    In addition to the above three aspects, chemical and mechanical

durabilities are also required for practical fiber applications. Highly hygroscopic material, even if it has very low-loss, is not preferable as a fiber material. Mechanical strength is also important factor for the material selection, because we bend and pull fiber cables severely. These additional aspects could be solved with some engineering developments using strength tension members and moisture proof film, however, it is preferable that fiber material itself has high chemical and mechanical durabilities.

From the historical viewpoint, many kinds of light-transmission waveguide were studied for their practical applications. Therefore, it is valuable to think about what kinds of materials, gas, liquid, crystal or glass, are most suitable for high quality optical fibers. Optical losses of various gases are very low in the visible and near infrared regions. Fatal defect of gaseous materials is the difficulty of their refractive-index control. Many trials to realize gas waveguide[1] have been done, in which the refractive index is controlled by changing the gas temperature between the center and periphery of a guide. Hollow waveguide,[2,3] which is a metallic pipe or a dielectric pipe, is sometimes used for infrared region. Transmission loss of hollow waveguide is strongly dependent on the pipe materials and the surface finish.

Some liquid materials are also low loss medium for optical fibers. Wide variety of refractive-index can be chosen in transparent liquid materials. Liquid core fiber,[4] liquid filled thin glass pipe, has been studied in the development history of low loss fibers. However, the refractive-index of liquid changes a considerable amount so that it is fairly hard to maintain the waveguide characteristics precisely. Furthermore, only step-index-type multimode fibers can be fabricated for practical use.

Solid state materials have stronger interaction with light wave than gases and liquids. Therefore, in general, the transmission loss may be higher, however, the optical properties are much more stable compared to gases and liquids. Solid state materials have strong absorption at ultraviolet wavelength and infrared wavelength regions, which are attributed to the electronic transition and molecular vibrations, respectively. These absorption peaks affect the absorption loss at the visible and near-infrared wavelength region. In order to find low-loss materials, it will be better to select materials whose strong absorption wavelengths are far from each other. Solid-state materials are classified in two groups: crystalline state and amorphous state. Optical characteristics of the two states are quite different, even if the atomic composition is the same. Crystalline materials, such as quartz, sapphire, and alkaline halides have essentially very low loss compared to amorphous materials. Because Rayleigh scattering loss, which originates from the density fluctuation, is less than one tenth of that of amorphous materials. It can be said that the best materials may be found in single crystals from the viewpoint of intrinsic loss. High speed crystal growth or mass-production of single crystal fiber, however, is very difficult and moreover the refractive-index control of crystalline materials are also difficult. Indeed, high doping of refractive-index modifier to the host crystal sometimes makes it difficult to grow a large single crystal. Furthermore, as-grown single crystal have specific crystal surfaces, this leads to roughness on the surface and core/clad interface of the fibers, which introduce a large scattering loss.

Although the transmission loss is comparatively high, polycrystalline materials are usable. The major loss increase in polycrystalline materials is caused by light scattering at the grain boundary. The scattering intensity is very strong. Therefore, the transmission loss of polycrystalline fiber is larger than expected from the intrinsic material property. In the infrared wavelength region, polycrystal fibers are used, because there are few suitable materials other than alkali-halide crystals available.

Noncrystalline or amorphous materials are ideal for optical fibers, because some oxide glass, fluoride glass and chalcogenide glass are highly transparent in visible and infrared wavelength regions. These glasses permit comparatively easy fabrication of fibers without introducing extrinsic scattering sources such as specific crystal surface roughness or grain boundaries. Moreover, the refractive-indexes of these glasses are easily controlled by changing the constituent ratio, which does not severely limit the glass formation condition. Rayleigh scattering loss, originating from the density and constituent fluctuations of amorphous materials, is larger than that of crystalline materials. However, the Rayleigh scattering loss is much smaller than the loss caused by grain boundary scattering in polycrystalline materials.

Glasses are made by cooling certain molten materials in such a manner that they do not crystallize but remain in an amorphous state, and their viscosity increases to such high values that they are solid. Materials having this ability to cool without crystallizing are relatively rare, silica, SiO_2, being the most common example. Some oxides and their mixtures can be made into glasses easily. Oxide glasses, especially silicate glass, have high transparency in the visible and near-infrared wavelength region. Chemical and mechanical durabilities of silicate glasses are also high. The refractive-index of silicate glass can be changed easily by doping with other oxides such as GeO_2, Al_2O_3 and TiO_2. Furthermore, the raw materials are relatively cheap. Therefore, silica glass is one of the most suitable materials for low loss optical fibers.

The idea to use silica glass for optical communication is very old, and some of them can be found in a Japanese patent[5] filed in 1936. Predictions based on scientific and technological investigation, however, were firstly published in 1965 by Kao and Hockham[6] of Standard Telecommunication Laboratories. They speculated that if the content of transition metals in silica glass was reduced to 1 ppm, then the transmission loss due to absorption would be reduced to below 20 dB/km. Scattering loss of silica glass was also estimated to be reduced to about 1 dB/km. Therefore, it was estimated that the total optical loss could be reduced to less than 20 dB/km.

In 1970, Kapron and their coworkers[7] of Corning Glass Works succeeded in the fabrication of a fiber with the low transmission loss of 20 dB/km. This success triggered explosive research activities toward the development of practical optical fibers. After the intensive development of various fabrication method and material research for about 10 years, very low-loss fibers of less than 0.2 dB/km have been attained using doped silica glasses. Although there are various reports discussing the possibility[8] of ultralow loss, less than 10^{-2} dB/km, for alkali halides, it does not seem to be easy to achieve low-loss fibers of less than 0.2 dB/km except with high silica fibers.

Oxide glasses, fluoride glasses, chalcogenide glasses, alkali-halide crystals and plastics are the representative candidates for low-loss optical fiber materials. In the following section, fundamental characteristics of the representative transparent materials in visible and near-infrared wavelength region will be reviewed from the viewpoint of optical fiber materials. Figure 1.1 shows the transparent wavelength regions, where the transmission loss is less than 1000 dB/km, for the representative transparent materials in the visible and near-infrared wavelength region.

1.1.2 Oxide glass

When a liquid material is cooled down gradually, it will change to crystal expelling the specific latent heat at the characteristic temperature, melting temperature. Some special materials do not crystallize below the melting temperature and their viscosity increase to such high value that they are solid

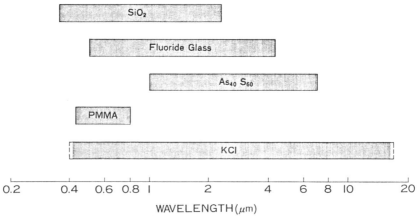

Fig. 1.1. Rough sketch of transparent regions for the representative highly transparent materials. The wavelength regions for the transmission loss less than 1000 dB/km are shown. For KCl crytical, the wavelengths are estimated by the extrapolation of the available data.

with disordered arrangement of atoms, quite like that in the liquid state. The super cooled liquid is called glass, and is unstable from the thermodynamic viewpoints, because the enthalpy of glassy state is higher than that of crystal.

Some oxides have this ability of forming glasses by themselves, which are named glass network former. SiO_2, GeO_2, B_2O_3, P_2O_5, As_2O_3, and Sb_2O_3 are known as glass network former. Network formers can form glass not only by themselves but also by the mixture with other network formers at any mixing ratio. The requirements to be satisfied to form stable galss were presented by Zachariasen,[9] that is the conditions to form a glass with energy comparable to that of the crystalline form: (1) an oxygen atom is not linked to more than two atoms; (2) the number of oxygen atoms surrounding atoms must be small; (3) the oxygen polyhedra share corners with each other, not edges or faces; (4) at least three corners in each oxygen polyhedron must be shared.

Some oxides, called network modifier, can not form a glass by themselves, but they can form a glass as a mixture with network formers to a certain content. Oxides of alkali metol and alkali-earth metal such as CaO, SrO, BaO, Li_2O, Na_2O, K_2O, Rb_2O, and Cs_2O, are typical network modifiers. Physical properties of the glass, refractive index, thermal expansion coefficient, fictive temperature etc., can be changed easily by adding other network formers or network modifiers. TiO_2, SnO_2, Al_2O_3, ZrO_2, Fe_2O_3, BeO, MgO, NiO, ZnO, CoO, FeO, and PbO have intermediate characteristics between network former and network modifier and are called intermediate oxides.

Most transition metal oxides in oxide glass introduce strong absorption in the visible and near-infrared wavelength region. Cr ions at a concentration of 1 ppb, for example, introduce a loss of about 1 dB/km at 1.0–1.8 μm. Therefore, transition metals should be removed from the fiber materials in order to make low loss optical fibers. Some metals, such as Cu, Ag, Au, and Pt, form colloids in glass. Au and Cu colloids make the glass red and Ag colloid makes the glass yellow. Some compounds of S, Se, Te and P also form colloids. Glass containing the mixture of CdS and CdSe changes color from dark red to yellow depending on the content ratio of CdS and CdSe. These transition elements and colloid-forming elements are not preferable for fiber materials, even if they are glass network formers.

Alkali metal oxides are sometimes used to decrease the glass softening temperature. Chemical durability of the glass containing alkali metal oxides is comparatively poor, because alkali ions in the glass dissolve in water. Alkali metal, alkali-earth metal and heavy metal atoms are very hard to reduce transition metal atoms to the contamination level less than 1 ppb. Therefore, it is not preferable to use alkali oxides for making low-loss optical fibers.

The kinds of oxides suitable for fabricating low-loss fibers are limited strongly by the reason mentioned above. Removing these undesirable atoms from periodic table, it becomes clear that only a few atoms are applicable for oxide glass for low-loss fibers as shown in Fig. 1.2. Furthermore, except for silicon oxide, glass network formers do not necessarily have strong stability against moisture and some of them are toxic. Therefore, it seems to be a reasonable choice from the practical viewpoint that silica glass is adopted as a low-loss material from among other oxides. The optical properties of silica glass will be given in Chapte 2.

1.1.3 Halide glass[10-14]

Some fluorides and chlorides can be glass formers. BeF_2, ZrF_4, HfF_4, and AlF_3–BaF_3 are typical fluoride glasses. These fluoride glasses are promising materials for fiber use in the infrared wavelength region. BeF_2, like SiO_2, essentially satisfies Zachariasen's glass forming conditions and easily forms glass on cooling from the molten state. The difference is that the ionic valence of BeF is one half of that of SiO_2. In spite of potentially attractive properties, few trials of making fibers with fluoroberyllate glass have been done. This is so because these materials are highly toxic and hygroscopic for practical handling. A family of glass of the heavy metal fluorides are considered as promising candidates although they are not stable against moisture compared with oxide glasses. These glasses have very low-loss in the wavelength region longer than 2 μm where the loss of oxide glass is comparatively high. The glasses generally contain 60–70 mol% ZrF_4 as the primary constituent and 20 mol% BaF_2 as a network modifyer with a small amount of another metal fluoride, like ThF_4, LaF_3 and GdF_3, playing the role of the glass stabilizer.

Chloride glass would be alternate candidate for infrared glass fiber, although they are too hygroscopic for practical use. $ZnCl_2$ and $BiCl_3$–KCl glasses have been reported.

Transition metal ions in halide glass system have also their particular absorption spectra. OH radicals and transition metal ions are major sources of absorption loss as in the case of oxide glasses.

1.1.4 Chalcogenide glass[15-26]

Calcogenide glasses, which are solid solution of metal sulphides, selenides and tellurides such as As_2S_3, As_2Te_3, GeS_3, Tl_2Se, Sb_2S_3 are available with a stable vitreous state and a wide transmission wavelength range. They have a large glass formation area, stability against moisture and longer cutoff wavelength compared to those of oxide and fluoride glasses.

1.1.5 Single crystals and polycrystals

Single crystals are an ideal material from the viewpoint of intrinsic transmission loss, because Rayleigh scattering of single crystal is generally one tenth as small as that of amorphous materials. Although the intrinsic absorption loss of alkali halide materials is estimated to be very small, they can not be obtained in the amorphous state. So, these materials must be used

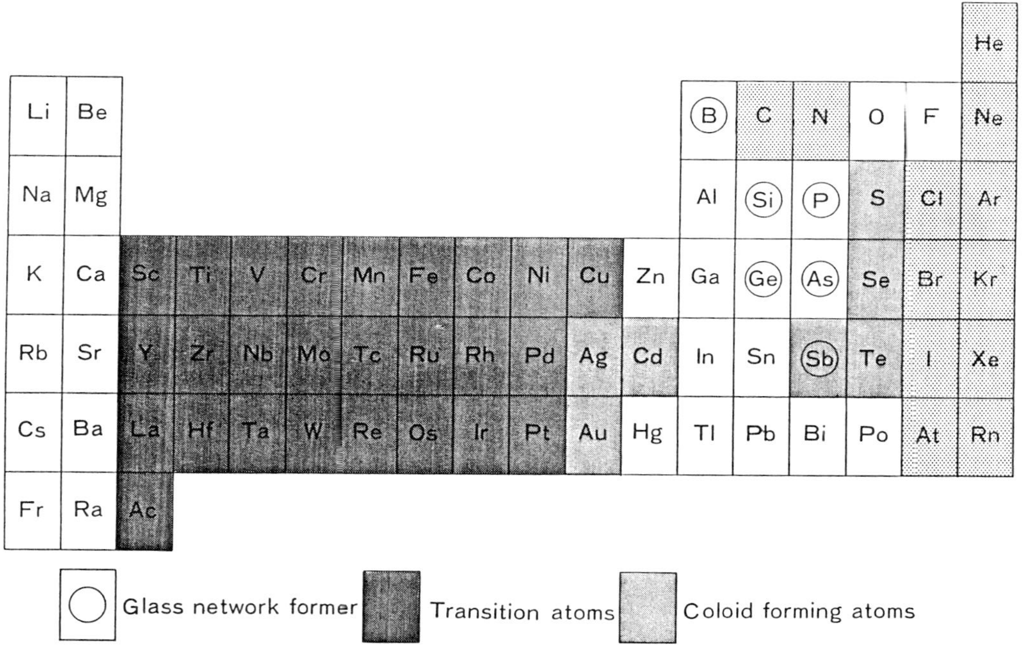

Fig. 1.2. The table of the elements whose oxides are glass network former, light absorbing materials (transition atoms and copper) and light scattering materials (colloid forming atoms). Gaseous oxides and elements are also shown in the table. This table shows that very small number of atoms are left for highly transparent oxide glasses.

in crystalline state. However, the fabrication techniques of single crystal fibers have not been well developed. There is a report[27] that sapphire filaments several hundred feet in length and diameter of 0.1–0.5 mm have been grown at speeds of 2.5–5 cm/min. This growth process is called EFG (edge-defined film-fed growth) method. It is difficult to increase the growth speed in this method. Furthermore, it is not easy to control the refractive-index and the cross sectional shape.

Recently, single crystal fibers[28-31] have been also grown from KRS-5 (TlBrI), AgBr, KCl, CsBr, and CsI. These fibers are unclad type or plastic loose clad type. The loss values of these fibers are far from the intrinsic absorption loss. Polycrystals have quite a great amount of grain boundary scattering loss, halide polycrystals are a logical choice since the fundametntal phonon absorption bands are located in the far infrared and the optical transparency of these materials covers a much broader spectral range than those of other materials.

1.1.6 Plastics[32]

Some kinds of plastics are highly transparent in the visible region. Typical plastics used for optical fibers are poly-methyl methacrylate (PMMA) and polystyrene, because these polymers are easily purified at the monomer level and no condensation reaction is involved in obtaining transparent polymer. The transmission loss of plastics is limited mainly by vibrational absorption due to carbon-hydrogen bond and by Rayleigh scattering. The strong absorption by C–H bond vibration and the harmonics is shown in Fig. 1.3. These absorption peaks can be shifted to longer wavelength region by substituting the hydrogen atoms with deuterium. Then, the total loss could be reduced. The estimated loss limit of PMMA and polystyrene is 10 dB/km.

1.2 Optical Transmission Loss of Fiber Materials

1.2.1 Introduction

The optical properties of solids are studied intensively as a powerful tool in our understanding of electronic and mechanical characteristics of solid as described in various textbooks of solid state physics. These investigations are related to the strong interaction between light and materials. Our interest, however, is how weak the interaction could be and there are few theoretical investigations about this viewpoint. Therefore, it is difficult to answer exactly the following questions: what is the most transparent solid state material? or how small can the light transmission loss in solids be? Table 1.1 shows the optical loss origins and the wavelength dependence described in various textbooks and papers. Figure 1.4 shows a schematic absorption loss spectrum over a wide wavelength region from milimeter wavelength to ultraviolet light. The spectrum was synthesized from data of As_2S_3.[33,34] Highly transparent

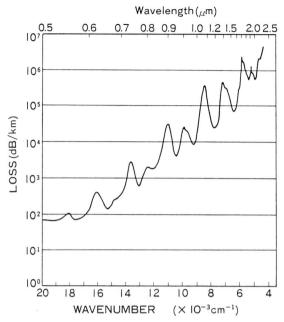

Fig. 1.3. Transmission loss property for polymethylmethacrylate (PMMA). [After Kitano et al.[1.32]]

materials generally have similar absorption loss spectra to the figure. Total loss spectrum of a material is generally modified from the absorption loss spectrum by adding scattering loss mainly in visible and near infrared wavelength regions.

Molecular vibration and electronic transition causes strong absorption at infrared and ultraviolet wavelength regions, respectively. These strong absorption bands are related to the intrinsic absorption loss at visible and near infrared regions where the lowest loss of a materials can be expected. Urbach's tail, weak absorption tail and multi-phonon absorption tail could be the loss origins of fiber materials. Figure 1.5 shows the loss spectrum[35] of a chalcogenide fiber, which has clear spectra of these three tails. Absorption loss caused by impurities such as OH-ions and transition metal ions has been reduced to very small level in case of high-silica glasses.

Rayleigh scattering, Brillouin scattering and Raman scattering are also intrinsic loss related to the density fluctuation and optical phonons. The dominant scattering loss origin is generally Rayleigh scattering and others are very weak except for stimulated scattering.

In this section, major loss origins of fiber materials at visible and near infrared wavelength regions will be mentioned. A guideline for selection of

Table 1.1. Intrinsic loss origins of solid state materials.

Spectrum		Electronic Transition Absorption				Scattering			Impurity	Free-carrier Absorption	Phonon Absorption		
		Inter Band	Intra Band	Exiton	Ulbach Tail	Rayleigh	Brillouin	Raman			Phonon	Multi Phonon	Fl Tail
		band	band	band	e^w	λ^{-4}	λ^{-4}	λ^{-4}	band	λ^2	band	e^{-w}	λ^2
Insulator	Crystal	S	None	S	S	W	W	W	S	W	S	S	S*
Insulator	Amorphous	S	None	S	S	S	W	W	S	W	S	S	S
Semi-Conductor	Crystal	S	S*	S	S	W	W	W	S	S	S*	S	S
Semi-Conductor	Amorphous	S	S*	S	S	S	W	W	S	S	S*	S	S
Metal		S					S		S	S	S	S	S

S: Strong, W: Weak, * Depend on the material

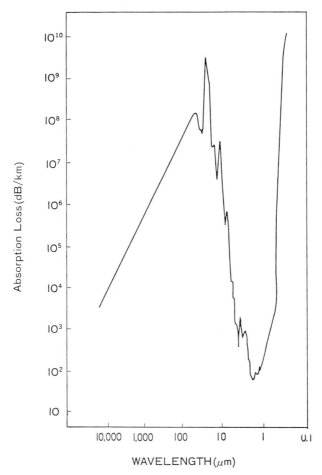

Fig. 1.4. Absorption loss spectrum for As_2S_3 over a wide wavelength region from milimeter wave to ultraviolet light. This spectrum is a typical one for highly transparent materials. [After Treacy et al.[1.33] and Kanamori[1.34]]

materials for low-loss fibers will be given from various experimental results of these major loss sources. The key points for reducing the loss of fiber materials and the loss limit of various fiber materials will be also found below.

1.2.2 Phonon absorption

In the infrared region, the absorption spectra are usually determined by phonon absorption, which is caused by bending and stretching vibration of constituent molecules. It has been recognized that multiphonon process

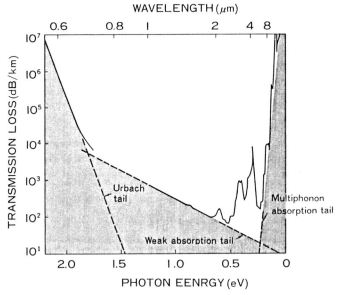

Fig. 1.5. Transmission loss versus photon energy for $As_{40}S_{60}$ bulk glass and unclad fiber. This spectrum shows a typical Ulbach tail and Weak absorption tail. [After Kanamori et al.[1.34]]

provides the intrinsic residual absorption in the shorter wavelength region than the phonon absorption wavelength.

Phonon absorption frequency will be given by a simple model. With crystals having two atoms per primitive cell, there are three acoustic branches and three optical branches to the phonon dispersion relation. Interaction between photon and phonon takes place at the cross point of the optical phonon branch and the dispersion relation of photon; $\omega = ck$. The momentum of a photon is negligible compared to phonon momentum, the cross point is at $k=0$. The vibrational frequency, ω_T, at the cross point is given by the following equation for transverse mode.

$$\omega_T^2 = 2C\left(\frac{1}{M_1} + \frac{1}{M_2}\right) \qquad (1.1)$$

where M_1, M_2 are the mass of atoms, and C is a force constant. The light frequency, ω_T, depends on the mass of two atoms and is absorbed to the crystal resonantly, the absorption coefficient at the frequency is very large. The frequency, ω_T, will be small when the mass of two atoms, M_1 and M_2, are large and when the binding force, C, between the two atoms is small.

We can observe weak absorption peaks at shorter wavelength, which can not be attributed to phonon absorption. It has been recognized that multiphonon process as provide these residual absorption[35-39] in the shorter wavelength region. Multiphonon absorption results from the combined effect of two interrelated mechanisms: the electric moment associated with the distorted charge density of real materials to which light couples directly; and the anharmonic interionic potential which stems from the interactions between electronic charge densities. Optically active modes excited via the interaction with electronic moment dissipate energy through anharmonic interaction. A detailed analysis and experimental results indicate that the frequency dependence of the absorption coefficient is given by

$$A_{MP} = K \exp(-\gamma \frac{\omega}{\omega_0}) \quad (1.2)$$

where K, γ, ω_0, are parameters characteristic of the material in question. The value of γ for various materials lies between 4 and 5, and the value, ω_0, is related[37] to the longitudinal optical phonon absorption wavenumber. Therefore, it is known from the equation that low-loss materials are composed with heavy atoms and strong binding forces between the atoms. Mass effects of constituent atoms are clearly observed in alkali-halide crystals and deuterated poly-methyl methacrylate (PMMA). CsI is transparent at a longer wavelength where KCl is not transparent. Deuterated PMMA, in which the hydrogen atoms of PMMA are substituted with deuterium, is transparent at longer wavelengths where regular PMMA has high transmission loss. Multiphonon absorption is related to the electric moment of the material, in other words, it is related to the ionicity of the atomic bonding. In case of ionic crystals, clear relation following to eq. (1.2) can be observed. When the ionicity of a crystal is weak, nonlinearity of the optical interaction becomes strong and the fine structure of absorption spectra can be observed. Figure 1.6 shows the multiphonon absorption tails for various materials, and Table 1.2[40] shows the values of K, γ and ω_0.

Transmission loss at longer wavelength regions than the fundamental phonon absorption peaks can be represented by an equation of the form[41,42]

$$n\alpha = K (\hbar \omega)^\beta \quad (1.3)$$

where α is the absorption coefficient, $\omega/2\pi$ is the frequency of the far infrared radiation, and n is the refractive index. For most of the materials investigated, $\beta < 2$. The origin of this tail is due to a disorder-induced coupling of the radiation to vibrational modes. The values[42] β for SiO_2, GeO_2, B_2O_3, As_2Se_3 and As_2S_3 are 2.0, the absorption loss decreases gradually with increasing wavelength. Therefore, low absorption loss can not be expected in the far infrared region. Absorption loss spectra for various materials are shown in Fig. 1.7.[42]

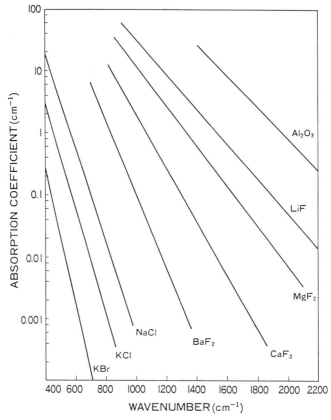

Fig. 1.6. Absorption coefficient for crystalline materials by multiphonon absorption. [After Deutsch[1.40]]

Table 1.2. Multiphonon absorption coefficients.

Materials	A(cm^{-1})	Eu	E_U (eV)	hw(eV)
KCl	1.26×10^{10}		0.033	7.76
KBr	2.4×10^{6}		0.033	
SnO$_2$	1.5×10^{5}		0.037	3.75
KI	5.9×10^{9}		0.032	5.89
NaCl	1.2×10^{10}		0.034	8.03
As$_{40}$S$_{60}$	—		0.054	—
As$_{38}$Ge$_{5}$Se$_{57}$	—		0.050	—
Ge$_{20}$S$_{80}$	—		0.073	—

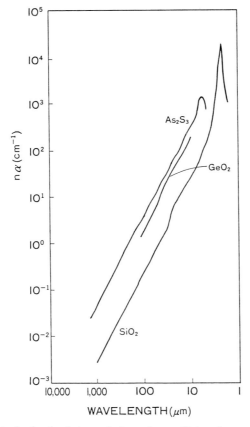

Fig. 1.7. Product of refractive-index and absorption coefficient for amorphous materials in far-infrared region. [After Pidgeon[1.42)]]

1.2.3 Electron transition absorption

Strong absorption has been observed in the ultra-violet region corresponding to the electron transition from valence band to conduction bend. Similar absorption could be observed for the photon energies just below the energy gap, which corresponds to the electron transition from the valence band to exciton levels.

Exponential absorption edges, similar to multi-phonon absorption edges, have been observed for photon energies smaller than the band gap energy, which is known as Urbach's rule.[43)] The absorption coefficient A_u varies with temperature T and photon energy $\hbar\omega$

$$\alpha = A_u \exp g(\hbar\omega - \hbar\omega_0) \tag{1.4}$$

where $\hbar$ is Planck's constant devided by 2π; k_B is Boltzmann's constant; A, ω_0, and g are fitting parameters. The empirical characterization was first enunciated to describe the observation in alkali-halide crystals. Not only alkali halides, II–IV compounds, III–V semiconductors and amorphous materials sometimes exhibit exponential tails. Table 1.3 shows the parameters of Urbach's rule for various materials.[44-48] The slope factor, E_U, is very small and is in the range 0.03–0.07 eV. Therefore, the absorption loss tail decreases abruptly at longer wavelength, and it does not affect the total loss of materials seriously.

1.2.4 Weak absorption tail

Does Urbach's rule persist at very low absorption region, where the absorption coefficient is smaller than 1 cm^{-1}? We can find another tail which is different from the Urbach tail but also follows an exponential law. The second absorption tail is called the weak absorption tail[49] and can be described as

$$\alpha = \alpha_0 \exp(\hbar \omega / E_U) . \quad (1.5)$$

The weak absorption tail was first observed in amorphous semiconductors such as As_2S_3, $Ge_{33}As_{12}Se_{55}$.[49] These tails have also been observed in alkali-halide crystals BaF_2 and SiO_2.[50] For most cases, the absorption constants follow Urbach's rule for $\alpha > 0.1$ cm^{-1} and E_U is in the range 0.03–0.07 eV. For $\alpha < 0.1$ cm^{-1}, one observes another exponential part of absorption where the slope factor, E_U, is 0.16–1.6 eV for various materials as is shown in Table 1.4. This tail could influence the total loss of materials at near infrared region where the scattering loss and phonon absorption loss are very small.

The origin of the weak absorption tail has not been fully clarified yet, it is thought to be the transition from localized states deep in the band gap to the

Table 1.3. Ulbach tail coefficients.

Materials	K(cm^{-1})	g	w$_0$
KCl	8696	4.2	213
KBr	6077	4.2	166
NaCl	24273	4.8	268
BaF$_2$	49641	4.5	344
CaF$_2$	105680	5.1	482
MgF$_2$	11213	4.4	617
LiF	21317	4.4	673
Al$_2$O$_3$	55222	5.0	900
NaF	41000	5.0	425

Table 1.4. Weak absorption tail coefficients.

Materials	E_U (eV)
LiF	0.55
KCl	0.53
NaCl	0.16 (78K)
CsI	0.35
BaF_2	0.87
KRS-6	0.28
SiO_2	0.41
As_2S_3	0.3
$As_{40}S_{60}$	0.28
$As_{38}Ge_5Se_{57}$	0.19
$Ge_{20}S_{80}$	0.39

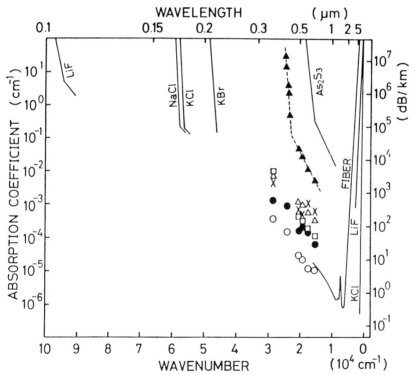

Fig. 1.8. Absorption coefficients for low-loss materials in visible region. Solid lines show reported absorption coefficients in ultra-violet and infrared region. ▲ KRS-6, × BaF_2, △ KCl, □ CsI, ● LiF, ○ SiO_2 (SUPRASIL-W1). [After Mori and Izawa[1.50]]

extended states. The localized states may be associated with deep potential fluctuations due to defects, disorder and impurity. According to experimental results, the absorption coefficient of alkali-halide crystals can be changed by thermal treatments[51] For example, for a KCl crystal heated up to about 600 K and then cooled down in a short time, quenched, the absorption coefficient increased. The quenched sample was again heated up to 600 K and cooled down slowly, annealed, then the coefficient returned to the initial value. The quenching and annealing effects have also been clearly observed in CsI crystals. It has also been observed that the absorption coefficient depends up on treatment in chloride gas. These experimental results support the idea that the origin of the weak absorption tail can not attributed to imputities but to the defects and dislocations. Weak absorption tail can be observed when the scattering loss, which has λ^{-4}-dependence, is comparatively low and the uv-absorption peak is seen to be at longer wavelength. Chalcogenide glasses are typical materials satisfying this condition. Figure 1.5 shows the loss spectrum of a chalcogenide glass.[34] In this figure, weak absorption tails are clearly observed. Weak absorption tails can be observed in various materials by laser calorimetric measurements as shown in Fig. 1.8. Therefore, this tail is very important to estimate loss-limit of ultra-low-loss materials.

1.2.5 Scattering loss[52-54]

The scattering loss is one of the major sources of loss in the visible and near infrared region. It is well known that there are three kinds of scattering, Rayleigh scattering, Brillouin scattering and Raman scattering. These are caused by density and concentration fluctuation, accoustic phonon, and optical phonon, interactions, respectively.

Light scattering in liquids and glasses is known to be due to microscopic variations in the local dielectirc constant associated with the random molecular structure of these materials. The spectrum and intensity of the light scattered through an angle θ, when a glass is illuminated with monochromatic light polarized normal to the scattering plane, are considered. The intensity per unit solid angle is given by

$$I = I_0 \, (\pi^2/\lambda_0^4) \, V^2 \, (1 + \cos \theta^2) \langle \Delta\varepsilon_k^2 \rangle \tag{1.6}$$

where I_0 and λ_0 are, respectively, the intensity and wavelength of the incident light, $\Delta\varepsilon_k$ is the kth Fourier component of the fluctuation in the dielectric susceptibility. In a single component substance, the mean square fluctuation in susceptibility, $\langle \Delta\varepsilon_k \rangle$, can be written in terms of fluctuation in density, $\langle \Delta\rho_k \rangle$

$$\langle \Delta\varepsilon_k^2 \rangle \approx (\partial\varepsilon/\partial\rho) \langle \Delta\rho_k^2 \rangle \,. \tag{1.7}$$

The mean-square density fluctuation can be calculated from the thermodynamic properties of the medium. For a glass,

$$<\Delta\rho_k^2> = \frac{\rho_0^2}{V} \{kT_f(\beta_T - \beta_s) + kT_f[\beta_s - (\rho \cdot v_\infty^2)^{-1}] + kT(\rho_0 v_\infty^2)^{-1}\} \quad (1.8)$$

where ρ_0 is the density, T_f is the fictive temperatue, that is the temperature at which the thermodynamic density fluctuations in melt are kinetically frozen into the glass. β_T and β_s are, respectively, the equilibrium isothermal and adiabatic compressibilities of the melt at T_f, and μ_∞ is the high-frequency sound velocity in the glass. Then the scattered intensity, I, is given by

$$I = I_0 (\pi^2/\lambda_0^4) V (1 + \cos^2\theta)(\rho_0 \frac{\partial \varepsilon}{\partial \rho})^2 \{kT[\beta_T - (\rho_0 v_\infty^2)^{-1}] + kT(\rho_0 v_\infty^2)^{-1}\}. \quad (1.9)$$

The first term, defined as I_R, represents the intensity appearing in Rayleigh lines and the second term, defined as I_B, represents the intensity of Brillouin lines. In a multicomponent glass sytem, eq. (1.9) must be modified to account for local fluctuations in composition. In a binary system, the scattering from composition fluctuations appears in the Rayleigh line in the spectrum, and the intensity is given by

$$I_R^C = I_0 \frac{\pi^2}{\lambda^4} V^2 (1 + \cos^2\theta)(\frac{\partial \varepsilon}{\partial C})^2 \frac{kT_f'}{N'} (\frac{\partial \mu}{\partial C})^{-1} \quad (1.10)$$

where C is the concentration defined as the ratio of the number of moles of solute N to the number of moles of solvent N', μ is chemical potential

The first term and the second term of eq. (1.9) can be rewritten using the photoelastic (Pockels) coefficient, p_{12}, for an isotropic solid.

$$I_R = I_0 \frac{\pi^2}{\lambda_0^4} V (1 + \cos^2\theta)(n^4 P_{12})^2 k T_f [\beta_T - (\rho_0 v_\infty^2)^{-1}] \quad (1.11)$$

$$I_B = I_0 \frac{\pi^2}{\lambda_0^4} V (1 + \cos^2\theta)(n^4 P_{12})^2 k T (\rho_0 v_\infty^2)^{-1} \quad (1.12)$$

where n is the refractive-index of the material. It should be noted that these expressions have the explicit λ^{-4}-dependence, indicating that the scattering loss substantially decrease at longer wavelength. It should also be noted that the intensity of Rayleigh scattering increases proportionally to the fictive temperature.

The intensity ration of Rayleigh and Brillouin scattering, I_R/I_B, which is known as Landau-Placzek ratio, is 10–45 for various glasses and is 23.2 for fused silica glass. The density fluctuation in crystals is comparatively small, therefore, the Brillouin scattering is dominant in high quality crystals. Figure 2.9 shows the scattering spectra of fused silica and silica crystal. The intensity of synthesized silica crystal was one-fifteenth that of synthesized fused silica.

Raman scattering has also λ^{-4}-dependence. The intensity of Raman scattering is usually very weak except for stimulated scattering, and it is smaller than 10^{-3} of Rayleigh scattering.

REFERENCES

1) D. W. Berreman, "A lens or light guide using convectively distorted thermal gradients in gasses", *Bell Syst. Tech. J.*, Vol. 43, p. 1469 (1964).
2) C. C. Eaglesfield, "Optical pipeline, a tentative assessment", *Proc. IEE*, Vol. 109B, p. 26 (1962).
3) E. A. J. Marcatili and R. A. Schmeltzer, "Hollow metallic and dielectric waveguides for long-distance optical transmission and lasers", *Bell Syst. Tech. J.*, Vol. 43, p. 1783 (1964).
4) J. Stone, "Optical transmission loss in liquid-core hollow fibers", *IEEE J. Quantum Electron.*, Vol. QE-8, p. 386 (1972).
5) H. Negishi and T. Seki, Jpn. Patent 125,946 (1938).
6) K. C. Kao and G. A. Hockman, "Dielectric fiber surface waveguide for optical frequencies", *Proc. IEE*, Vol. 113, p. 1151 (1966).
7) F. P. Kapron, D. B. Keck and R. D. Maurer, "Radiation losses in glass optical waveguides", *Tech. Digest Conf. on Trunk Telecom. by Guided Waves*, p. 148 (1970).
8) C. H. L. Goodman, "Devices and meaterials for 4 μm band fiber optical communication", *IEEE J. Solid-State Circuits*, Vol. 2, p. 129 (1978).
9) W. H. Zacharisen, "The atomic arrangement in glass", *J. Am. Chem. Soc.*, Vol. 54, p. 3841 (1932).
10) S. Mitachi and T. Manabe, "Fluoride glass fiber for infrared transmission", *Jpn. J. Appl. Phys.*, Vol. 19, p. L313 (1980).
11) M. G. Drexhange, B. Bendow, T. J. Lorentz, J. Mansfield and C. T. Moynihan, "Preparation of multicomponent fluoride glass fibers by single crucible technique", *Tech. Dig. 3rd Int. Conf. on Int. Optics and Optical Fiber Commnication*, p. 32 (1981).
12) R. J. Ginther and D. C. Tran, "Fluoride glasses for IR transmitting fibers", *Tech. Dig. 3rd Int. Conf. on Int. Optics and Optical Fiber Communication*, p. 32 (1981).
13) S. Mitachi, T. Miyashita and T. Manabe, "Preparation of fluoride optical fibers for transmission in the mid-infrared", *Phys. Chem. Glasses*, Vol. 23, p. 196 (1982).
14) L. G. Van Uitert and S. H. Wemple, "$ZnCl_2$ glass: A potential ultralow-loss optical fiber material", *Appl. Phys. Lett.*, Vol. 23, p. 57 (1978).
15) N. S. Kapany and R. J. Simms, "Recent developments of infrared fiber optics", *Infrared Phys.*, Vol. 5, p. 69 (1965).
16) S. Shibata, Y. Terunuma and T. Manabe, "Ge-P-S chalcogenide glass fibers", *Jpn. J. Appl. Phys.*, Vol. 19, p. L609 (1980).
17) S. Shibata, Y. Terunuma and T. Manabe, "Sulfide glass fibers for infrared transmission", *Mater. Res. Bull.*, Vol. 16, p. 703 (1981).
18) C. Le Sergent, "Infrared glass optical fibers for 2 toll micrometer band", *Advances in IR fibers II*, Tech. Dig. SPIE, paper 320-03 (1982).
19) E. M. Dianov, "Materials for infrared low-loss fibers", *Advances in IR Fibers II*, Tech. Dig. SPIE, paper 320-04 (1982).
20) A. Bornstein, N. Croitoru and E. Maron, "Chalcogenide infrared glass fibers", *Advances in IR Fiber II*, Tech. Dig. SPIE, Paper 320-18 (1982).

21) T. Miyashita and Y. Tarunuma, "Optical transmission loss of As–S glass fiber in 1.0–5.5 μm wavelength region", *Jpn. J. Appl. Phys.*, Vol. 21, p. L75 (1982).
22) J. A. Savage and S. Neilsen, "Chalcogenide glasses transmitting in the infrared between 1 and 20 μm—A state of the art review", *Infrared Phys.*, Vol. 5, p. 195 (1965).
23) A. R. Hilton and C. E. Jones, "Non-oxide IVA-VA-VIA chalcogenide glass. Part 2. Infrared absorption by oxide impurities", *Phys. Chem. Glasses*, Vol. 7, p. 112 (1966).
24) P. A. Young, "Optical properties of vitreous arsenic trisulfide", *J. Phys. C: Solid St. Phys.*, Vol. 4, p. 93 (1971).
25) D. L. Wood and J. Tauc, "Weak absorption tails in amorphous semiconductors", *Phys. Rev.*, Vol. B5, p. 3144 (1972).
26) C. T. Moynihan, P. B. Macedo, M. S. Maklad, R. K. Mohr and R. E. Howard, "Intrinsic and impurity infrared absorption in As_2Se_3 glass", *J. Non-Cryst. Solids*, Vol. 17, p. 369 (1975).
27) H. E. LaBelle, Jr. and A. I. Mlavsky, "Growth of controlled profile crystals from the melt: part I—sapphire filaments", *Mater. Res. Bull.*, Vol. 6, p. 571 (1971).
28) G. Tangonan, A. G. Pastor and R. C. Pastor, "Single crystal KCl fibers for 10.6 μm integrated optics", *Appl. Opt.*, Vol. 12, p. 1110 (1980).
29) T. J. Bridges, J. S. Hasiak and A. R. Strand, "Single-crystal AgBr infrared optical fibers", *Opt. Lett.*, Vol. 5, p. 85 (1980).
30) Y. Mimura, Y. Okamura, Y. Komazawa and C. Ota, "Growth of fiber crystals for infrared optical waveguides", *Jpn. J. Appl. Phys.*, Vol. 9, p. L269 (1980).
31) Y. Okamura, Y. Mimura, Y. Komazawa and C. Ota, "CsI crystalline fiber for infrared transmission", *Jpn. J. Appl. Phys.*, Vol. 19, p. L649 (1980).
32) T. Kaino, M. Fujiki and K. Jinguji, "Preparation of plastic optical fiber", *Rev. ECL*, Vol. 32, p. 478 (1984).
33) D. Treacy and P. C. Taylor, "Multiphonon absorption in the chalcoganide glasses As_2S_3 and GeS_2", in *Optical Properties of Highly Transparent Solid*, edited by S. S. Mitra and B. Bendow, Plenum Press, p. 261 (1975).
34) T. Kanamori, Y. Terunuma and T. Miyashita, "Prepareation of chalcogenide optical fiber", *Rev. ECL*, Vol. 32, p. 469 (1984).
35) S. S. Mitra and B. Bendow, "*Optical Properties of Highly Transparent Solids*", Plenum Press (1975).
36) M. Hass and B. Bendow, "Residual absorption in infrared materials", *Appl. Opt.*, Vol. 16, p. 2882 (1977).
37) L. L. Boyer, J. A. Harrington, M. Hass and H. B. Rosenstock, "Multiphonon absorption in ionic crystals", *Phys. Rev. B*, Vol. 11, p. 1665 (1975).
38) M. Sparks, "Infrared absorption by the higher-order-dipole-moment mechanism", *Phys. Rev. B*, Vol. 10, p. 2581 (1974).
39) B. Bendow, H. G. Lipson and S. P. Yukon, "Residual lattice absorption in semiconducting crystals: frequency and temperature dependence", *Appl. Opt.*, Vol. 16 (1977).
40) T. F. Deutsch, "Absorption coefficient of infrared laser window materials", *J. Phys. & Chem. Solids*, Vol. 34, p. 2091 (1973).
41) U. Strom, J. R. Hendrickson, R. J. Wagner and P. C. Taylor, "Disorder-induced far infrared absorption in amorphous materials", *Solid State Commun.*, Vol. 15, p. 1871 (1974).
42) C. R. Pidgeon, G. D. Holah, F. Al-Berkdar, P. C. Taylor and U. S. Strom, "Application of submillimeter waveguide laser to the study of absorption in elemental amorphous solids", *Infrared Phys.*, Vol. 18, p. 923 (1978).
43) J. D. Dow and D. Redfield, "Toward a unified theory of Urbach's rule and exponential absorption edges", *Phys. Rev. B*, Vol. 5, p. 594 (1972).
44) T. Toniki, "Optical constants and exciton states in KCl single crystals II—the spectra and reflectivity and absorption constant", *J. Phys. Soc. Jpn.*, Vol. 23, p. 1280 (1967).

45) W. Martienssen, "Uber die Excitonenbanden der Alkalihalogenidkristalle", *J. Phys. & Chem. Solids*, Vol. 2, p. 257 (1957).
46) M. Nagasawa and S. Shionoya, "Urbach's rule exhibited in SnO_2", *Solid State Commun.*, Vol. 7, p. 1731 (1967).
47) U. Haupt, "Uber Temperaturabhangigkeit und Form der langwellingsten Excitonenbande in KI-Kristallen", *Z. Phys.*, Vol. 157, p. 232 (1959).
48) T. Miyata and T. Tomiki, "The Urbach tail and reflection spectra of NaCl single crystal", *J. Phys. Soc. Jpn.*, Vol. 22, p. 209 (1967).
49) D. L. Wood and J. Tauc, "Weak absorption tail in amorphous semiconductors", *Phys. Rev. B*, Vol. 5, p. 3144 (1972).
50) H. Mori and T. Izawa, "A new loss mechanism in ultra-low-loss optical fiber materials", *J. Appl. Phys.*, Vol. 51, p. 2270 (1980).
51) H. Mori, private communication.
52) J. Schroeder, R. Mohr, P. B. Macedo and C. J. Montrose, "Rayleigh and Brillouin scattering in K_2O-SiO_2 glasses", *J. Am. Ceram. Soc.*, Vol. 56, p. 510 (1973).
53) M. E. Lines, "Scattering losses in optical fiber materials. I A new parameterization", *J. Appl. Phys.*, Vol. 55, p. 4052 (1984).
54) M. E. Lines, "Scattering loss in optical fiber materials. II Numerical estimate", *J. Appl. Phys.*, Vol. 55, p. 4058 (1984).

Chapter 2

OPTICAL PROPERTIES OF PURE AND DOPED SILICA

Pure or doped silica glass is one of the best material for optical fibers from the viewpoint of making fiber with high quality transmission characteristics and of practical application for communication. In this chapter, details of loss characteristics, refractive-index and thermal properties of pure and doped silica glass will be given. These optical properties will be useful for designing fiber structure and for the estimation of transmission characteristics at various ambient conditions.

2.1 Transmission Loss

2.1.1 Introduction

Pure and doped silica glasses are highly transparent in the wavelength region from 0.4–2.0 μm. Transmission loss of silica glass fibers is caused by two reasons; one is intrinsic to the glass such as multiphonon absorption, Rayleigh scattering and UV absorption tail. The other is extrinsic, introduced in the fiber fabrication process such as structual imperfection and impurity absorption. Most of these extrinsic losses have been reduced by the recent progress in fabrication processes as low as to make the extrinsic losses undetectable except for OH ions absorption. Therefore, the transmission loss of silica fibers is mainly limited by Rayleigh scattering at shorter wavelengths and by multiphonon absorption at longer wavelengths as is shown in Fig. 2.1.[1] The UV absorption tail is the major loss origin in the wavelength region less than 0.4 μm. In this section, the loss origin of silica fibers and the dopant effect to the loss characteristics will be given.

2.1.2 Phonon absorption and multiphonon absorption
A) SiO_2[2]

The atomic arrangement of silica glass is thought to be random arrangement of SiO_4 tetrahedron. Figure 2.2 shows two-dimensional pictures of (a) crystal silica and (b) silica glass, the black dots show silicon atoms and

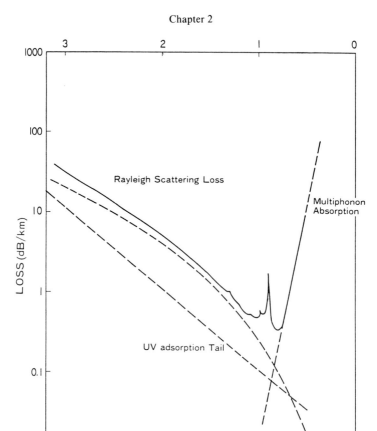

Fig. 2.1. Transmission loss spectrum of a silica glass fiber from visible to near-infrared region. The loss minimum is decided with Rayleigh scattering loss and multiphonon absorption loss characteristics of the glass. [After Osanai et al.[2.1]]

white dots show oxygen atoms. In reality, silicon atoms have one more bond to an oxygen atom and form three dimensional structure. The phonon absorption bands of nondoped silica correspond to the fundamental vibrations of a tetrahedral SiO_4 molecule as shown in Fig. 2.3. Figure 2.4 shows the absorption spectrum of pure silica glass. It has strong absorption bands at 9.1 μm (v_3), 12.5 μm (v_1) and 21 μm (v_4). The v_2 vibration mode shows absorption at 36.4 μm. The intensities of these absorption peaks are about 1×10^{10} dB/km. Nondoped silica has a number of absorption bands in the 3–8 μm region, which can be attributed to the overtones and combination tones of these

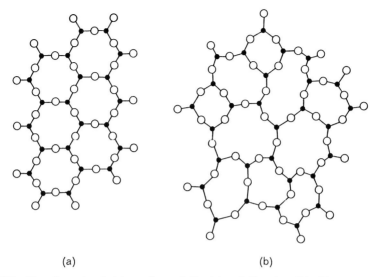

Fig. 2.2. Two-dimensional pictures of crystal silica (a), and silica glass: ● is silicon atom and ○ is oxygen atoms.

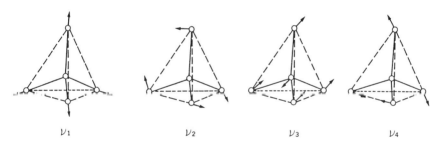

Fig. 2.3. Fundamental vibrations of a tetrahedral SiO_2 molecule. [After Izawa et al.[2.2]]

fundamental vibrations. The assignment of higher overtones and combination tones is rather uncertain for several reasons: (1) Because of the anharmonicity, the overtones of the triply degenerated infrared active vibration split into a number of subbands. (2) Since v_1+v_2 has approximately the same frequency as v_3, the combinations $n(v_2+v_2)$ are close to nv_3. (3) These centers of the bands are often very ill defined, because of the overlapping discussed under (1) and (2). The assignment of these bands in Table 2.1[2] is not the only possible one and is merely suggested. The measured peak intensities of these bands are also given in Table 2.1. The multiphonon absorption coefficient of pure silica at shorter wavelength is roughly given[3] by

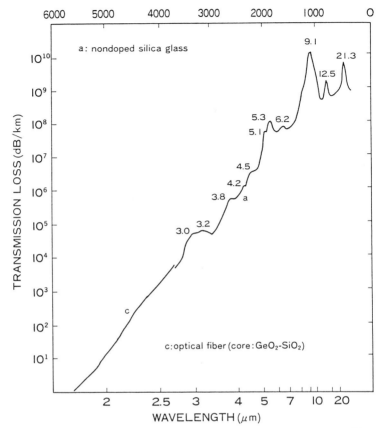

Fig. 2.4. Infrared absorption spectrum of pure silica glass. [After Shibata[2.6]]

$$A_{MP} = 7.81 \times 10^{11} \exp(-48.48/\lambda) \quad (dB/km) \tag{2.1}$$

where λ is the wavelength.

B) GeO_2

GeO_2 glass has an atomic structure similar to that of silica glass and has a strong absorption band at 11.4 μm which corresponds to the v_3 vibration mode. However, in GeO_2-doped fused silica, the spectrum is not a simple superposition of that of SiO_2 and GeO_2, but broader and more diffuse. The spectrum of 10 wt% GeO_2-doped fused silica was almost the same as that of nondoped silica except for a slight shift of absorption peaks to longer wavelength.

C) P_2O_5

In case of P_2O_5-doped fused silica, a characteristic absorption band was

Table 2.1. Wavelength and intensities of absorption bands in fused silica. [After Izawa[(2.2)]]

Dopant	Wavelength (μm)	Intensity (dB/km)	Identification[a]
None	21.3	5×10^9	ν_4
	12.5	2×10^9	ν_1
	9.1	1×10^{10}	ν_3
	6.2	8×10^7	$2\nu_1, \nu_3 + \nu_4,$
	5.3	1×10^8	$\nu_1 + \nu_3, \nu_2 + \nu_3 + \nu_3, 2\nu_1 + \nu_2$
	5.1	6×10^7	$3\nu_2 + \nu_3, \nu_1 + \nu_2 + 2\nu_4,$
	4.5	4×10^6	$2\nu_3, \nu_1 + \nu_2 + \nu_3,$
	4.2	1×10^6	$3\nu_1, \nu_1 + \nu_3 + \nu_4,$
	3.8	6×10^5	$2\nu_3 + \nu_4, 2\nu_1 + 2\nu_4,$
	3.2	6×10^4	$4\nu_1, 2\nu_3 + 2\nu_4, 2\nu_1 + \nu_3 + \nu_4$
	3.0	5×10^4	$3\nu_3, \nu_1 + \nu_2 + 2\nu_3,$
B_2O_3	15.4	6×10^8	ν_4'
	11.1	7×10^8	ν_3'
	7.4	1×10^9	ν_3''
	4.1	3×10^6	$\nu_3 + \nu_3'', 3\nu_1''$
	3.7	4×10^6	$2\nu_3'', 2\nu_1'' + \nu_2'' + \nu_3''$
	3.15	1×10^5	$2\nu_3'' + \nu_4, \nu_1 + \nu_3 + \nu_3''$
P_2O_5	3.8	1×10^6	P=O bond stretching vibration

[a] ν_4': B—O—Si bond bending vibrations.
ν_3': B—O—Si nonsymmetric bond stretching vibrations.
ν_3'': B—O—B nonsymmetric bond stretching vibrations.

observed at 3.8 μm which can be attributed to the first overtone of the P=O bond stretching vibration. Figure 2.5[2)] shows the absorption spectrum of P_2O_5 doped silica glass. Though fundamental vibration of the P=O (7.8 μm), P–O–P (8.7, 10.5, and 12.8 μm), O–P=O (15.4 μm) and O–P–O (21.1 μm) are known to exist in the highly doped P_2O_5–SiO_2 glass, these peaks are obscured by the SiO_2 absorption bands in 7 wt% P_2O_5-doped silica.

D) B_2O_3

In the case of B_2O_3-doped silica, the absorption spectrum is strongly affected as shown in Fig. 2.5. The fundamental vibrations of B–O bond are observed at 7.9, 12.4, 13.9, and 15.3 μm. The strongest absorption band, at 7.9 μm, which has been assigned to the bond stretch vibration of the boron sublattice against the oxygen sublattice, shifts to longer wavelength with decreasing B_2O_3 concentration in B_2O_3–SiO_2 glass. The absorption band at 3.7 μm, which can be reasonably assigned to the first overtone of the 7.4 μm band, is also strong and affects the absorption loss of the glass below this wavelength.

Based on the above measurements, we can conclude that the intrinsic infrared absorption loss increases by several orders of magnitude beyond the

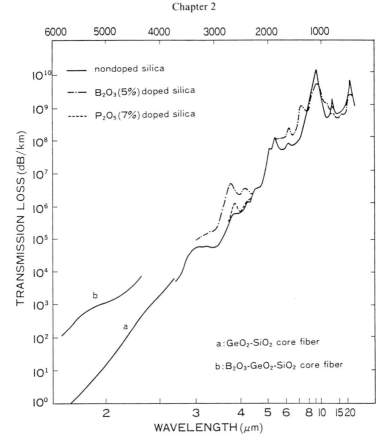

Fig. 2.5. Infrared absorption spectra of B_2O_3-doped silica glass and P_2O_5-doped silica glass. [After Izawa et al.[(2.2)]]

wavelength region 1.8–2.0 μm in doped fused silica fibers and that the infrared loss spectrum is affected more strongly by B_2O_3 compared to other dopants. Figure 2.6 shows the loss spectra of various fibers made of GeO_2, P_2O_5 and B_2O_3–GeO_2 doped glasses for the core. Multiphonon absorption tail shift to shorter wavelength has been clearly observed in the B_2O_3 doped silica core fiber.

2.1.3 Ultraviolet absorption and the tail

In the ultra-violet region, four strong absorption bands[4] are observed in silica as is shown in Fig. 2.7. The obsorption peaks at 10.2, 14.0 and 17.3 eV are attributed to excitonic transition while the transition at 11.7 eV is attributed to the conduction band. The band gap of silica is 8.9 eV which gives

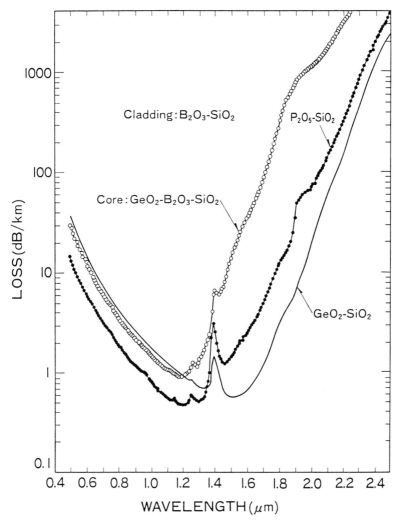

Fig. 2.6. Loss spectra of various fibers made of GeO_2, P_2O_5 and GeO_2-P_2O_5-doped silica glasses for the core. [After Osanai et al.[2.1]]

a rough upper limit for transmission at about 1400Å. The obsorption peaks for GeO_2 and B_2O_3 are 1650 Å and 1550 Å, respectively.

In case of silica glass, the electronic absorption tail has very small contribution in deciding the transmission loss, because the Rayleigh scattering loss is stronger than the tail in a wavelength range longer than 0.6 μm. Although there is no clear observation of Urbach's tail and weak absorption

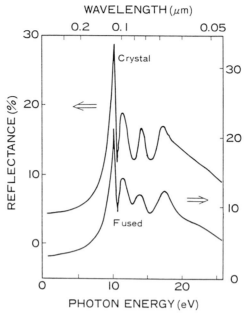

Fig. 2.7. Ultra-violet reflection spectra for silica glass. [After H. R. Philip, Optical transitions incrystalline and fused qualtz, *Solid State Commun.*, Vol. 4, p. 73 (1966)]

tail, it is reasonable that the observed UV-absorption tail at around 0.4 μm is a weak absorption tail. Because the Urbach's tail slope factor E_U is generally in the range of 0.03–0.07 eV and the weak absorption tail factor E_U is 0.16–1.6 eV for various materials, on the other hand, the observed slope-factor for silica glass is about 0.5 eV. When the Rayleigh scattering could be reduced by decreasing the softening temperature or by using single crystal, the weak absorption tail could affect the loss characteristics.

The absorption tails of doped silica glass are changed by the doping element and the quantity. Because the UV absorption peaks are at longer wavelength than that of silica. Figure 2.8 shows the absorption loss spectra for GeO_2, P_2O_5 and B_2O_3 doped silica glasses in the 0.2–0.5 μm wavelength region. The absorption peak at 0.24 μm is attributed to oxygen defects in silica glass. The peak could be decreased by adding OH ions of about 500 ppm in case of GeO_2-doped silica.

The tail of GeO_2-doped silica shifts toward longer wavelength region in proportion to the GeO_2 concentration, because the UV absorption peaks shift toward longer wavelength region as compared with pure silica and 0.20 and 0.25 μm absorption peaks are attributed to Ge^{4+} ions. The absorption coefficient of GeO_2-doped silica has been expressed by

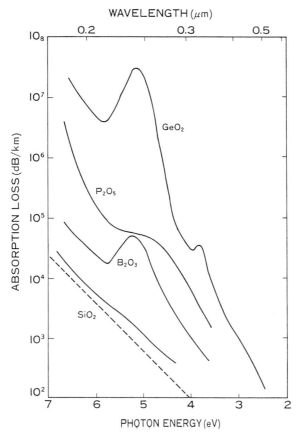

Fig. 2.8. Absorption loss spectra for GeO_2, P_2O_5 and B_2O_3-doped silica glass in the 0.2–0.5 μm wavelength region. [After Shibata[2.6]]

$$A_{UV} = 1.474 \times 10^{-11} \exp(E/0.268) \quad (dB/km/ppbw) \quad (2.2)$$

$$= 1542 \, \Delta / (446 \, \Delta + 6000) \times 10^{-2} \exp(4.63/\lambda) \quad (dB/km) \quad (2.3)$$

where E is photon energy and Δ is refractive-index difference of doped silica and pure silica.

2.1.4 Impurity absorption
A) Transition metal

Transition metal ions in silica glass give rise to broad, intense absorption in the visible and near-infrared region as shown in Fig. 2.9. The quantity of

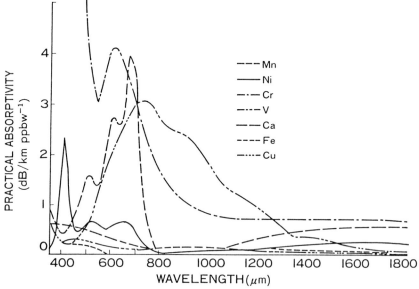

Fig. 2.9. Absorption due to transition metal ions in silica glass. [After P. C. Schultz, Optical absorption of the transition elements in vitreous silica, *J. Am. Ceram. Soc.*, Vol. 57, p. 309 (1974)]

each transition metal element giving 1 dB/km absorption loss is 2 ppb for Cr, 10 ppb for Mn, 20 ppb for Fe, 0.2 ppb for Co, and 2 ppb for Ni and for Cu. In the first stage of developing low-loss fibers, the transition metal ions were major loss origins. However, by using the "soot process", it is no longer a serious problem to reduce the transition metal impurities. In recent low-loss glasses, the absorption due to these ions can not be detected. The major reason of the success is that halides of Si, Ge, P, and B are used as the raw materials of glass. The halides of transition metal are easily removed from the raw materials of glass by a simple distilation.

B) *Hydroxyl ions*

The hydroxyl (OH) ion in silica glass gives rise to strong absorption[5] at 2.72 μm, which is assigned to fundamental stretching vibrational mode. The overtones and the combination tones with Si–O–Si bending vibration mode are observed at 2.20, 1.90, 1.38, 1.24, 1.13, 0.95, and 0.88 μm as shown in Table 2.2. The OH ions of 1 ppm give rise to the absorption loss of 65 dB/km at 1.38 μm, therefore, the OH ions should be reduced to the level lower than 1 ppb when the fiber is used at 1.0–1.6 μm region. The absorption peaks shift slightly by doping with other oxides. OH ions in GeO_2 glass give rise to an absorption peak at 2.86 μm as shown in Fig. 2.10[6]. In case of GeO_2-doped silica glass, the

Table 2.2. Absorption bands due to OH-ions and absorption loss caused by 1 ppm OH-ions. [After Kaiser[2.4]]

Wavelength (μm)	Frequrncy	Loss (dB/km)
0.60	$5\nu_3'$	0.006
0.64	$2\nu_1 + 4\nu_3'$	0.001
0.68	$\nu_1 + 4\nu_3'$	0.004
0.72	$4\nu_3'$	0.07
0.82	$2\nu_1 + 3\nu_3'$	0.004
0.88	$\nu_1 + 3\nu_3'$	0.09
0.945	$3\nu_3'$	1.0
1.13	$2\nu_1 + 2\nu_3'$	0.11
1.24	$\nu_1 + 2\nu_3'$	2.8
1.38	$2\nu_3'$	65.3
1.90	$2\nu_1 + \nu_3'$	10.3
2.22	$\nu_1 + \nu_3'$	260
2.72	ν_3'	10000

ν_1 : Si-O-Si symmetric vidration
ν_3' : O-H vibration

peaks shift toward longer wavelength with increasing GeO_2 content as is shown in Fig. 2.11[6].

2.1.5 Scattering loss

Rayleigh scattering loss is the major source of loss in the range of 0.6–1.6 μm. The scattering intensity is strongly affected by the doping element and the quantity because the fictive temperature decreases with doping, therefore, the density fluctuation will decrease. On the other hand, concentration fluctuation will increase with doping. Figure 2.12 shows the Rayleigh scattering loss[6] for various kinds of doped silica.

In case of GeO_2-doped silica glass, the scattering loss increases with increasing dopant quantity. The softening temperature decreases about 50°C with 10 mol% doping of GeO. Therefore, the concentration fluctuation seems to be a major effect in increasing the scattering loss. By doping 10 mol% of B_2O_3 or P_2O_5, the softening temperature decreases to about 1350°C. In the case of P_2O_5, the density fluctuation seems to decrease without adding strong concentration fluctuations. Therefore, the scattering loss of P_2O_5-doped silica glass decreases with increasing dopant quantity. In case of B_2O_5, the scattering loss increases by adding the dopant. The mechanism is not clear, but it seems

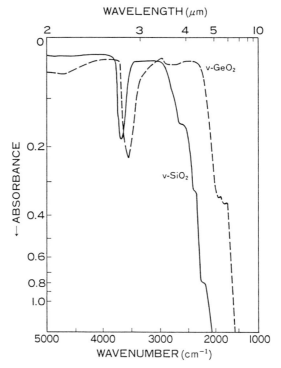

Fig. 2.10. Infrared absorption characteristics of SiO_2 glass and GeO_2 glass (the sample thickness is 2 mm). [After Shibata[2.6]]

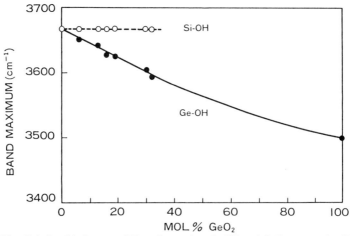

Fig. 2.11. Relationship between OH-peak wavelength shift and GeO_2 content in silica glass. [After Shibata[2.6]]

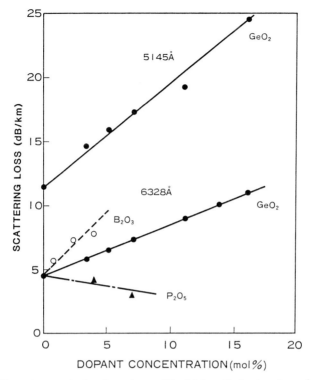

Fig. 2.12. Dopant concentration dependence of Rayleigh scattering loss for various kinds of doped silica. [After Shibata[2.6]]

to be related to the peculiar characteristics which has also been found in the refractive-index dependence on the doping quantity.

Rayleigh scattering coefficients for GeO_2-doped silica are shown as a function of refractive-index difference in Fig. 2.11. Using this data, we could formulate the Rayleigh scattering loss, A_{RS}, as a function of wavelength, λ, and refractive-index difference, Δ.

$$A_{RS} = (0.51\, \Delta + 0.76)/\lambda^4 \quad (dB/km). \qquad (2.4)$$

2.1.6 Total loss of high silica fibers

The wavelength dependence of total loss, A_T, can be formulated using experimental results. Here, we will describe the equation for GeO_2-doped single mode fibers. When the extrinsic losses are eliminated, the total loss, A_T, of GeO_2 doped singlemode fiber will be given by

$$A_T = A_{UV} + A_{RS} + A_{MP} .\tag{2.5}$$

Using eq. (2.1), (2.3) and (2.4), A_T is given by the following equation.

$$\begin{aligned}A_T =\ & 1542\,\Delta/(446\,\Delta + 6000) \times 10^{-2}\,\exp(4.63/\lambda) \\ & + \lambda^{-4}(0.51 + 0.76) + 7081 \times 10^{-11}\,\exp(-48.48/\lambda)\quad(\mathrm{dB/km}) .\end{aligned}\tag{2.6}$$

Figure 2.13 shows the GeO_2 concentration dependences of Rayleigh scattering loss, UV absorption tail and total loss as a function of relative refractive-index difference[3]. The loss minimum is given by

$$\frac{\partial A_T}{\partial \lambda} = 0 .\tag{2.7}$$

The calculated minimum loss is 0.18 dB/km at 1.55 μm. The refractive-index dependence of minimum loss wavelength and the loss values are given by Fig. 2.14 and Table 2.3[3]. These values stet be changed mainly by the reduction of scattering loss, however, the loss limit of high silica fibers is estimated to be around 0.1 dB/km.[7]

2.2 Refractive-Index

Refractive-index data for pure and doped silica glass are important for designing optical fiber structure such as optimum refractive-index profile of graded index fibers and cutoff wavelength of single mode fibers. For these

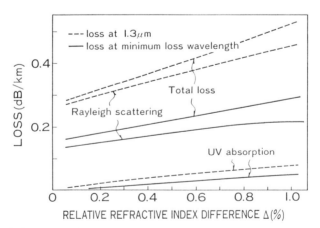

Fig. 2.13. GeO_2 concentration dependences of Rayleigh scattering loss, ultra-violet absorption tail and total loss at minimum loss wavelength and at 1.3 μm. [After Miya[2.3]]

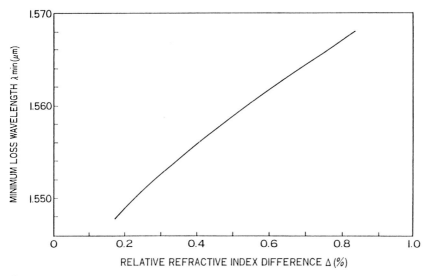

Fig. 2.14. Refractive-index or doping concentration dependence of minimum loss wavelength for GeO$_2$-doped silica core fibers. [After Miya[2.3]]

Table 2.3. Minimum loss, α_{min}, and minimum-loss wavelength, λ_{min}, for fibers with refractive-index differences, Δ. [After Miya[2.3]]

Δ (%)	0.2	0.3	0.4	0.5	0.6	0.7	0.8
α min(dB/km)	0.180	0.193	0.207	0.220	0.234	0.247	0.260
λ min(μm)	1.549	1.553	1.556	1.559	1.562	1.564	1.567

purpose, we require more detailed data such as dispersion, temperature dependence and photo-elastic effect. In this section, various experimental results of refractive-index measurement will be described.

2.2.1 Refractive-index dispersion

The refractive-index dispersion relation is given by the known equation

$$n^2 - 1 = \sum_{UV} \frac{4\pi N_0 f_{UV} e^2}{m(\omega_{UV}^2 - \omega^2)} + \sum_{IR} \frac{4\pi N_0 f_{IR} e^2}{M(\omega_{IR}^2 - \omega^2)} \qquad (2.8)$$

where m is the mass of atom, N_0 is number of atoms per unit volume, ω_{UV} and

ω_{IR} are the absorption band frequencies at ultra-violet and infrared region, respectively, f_{UV} and f_{IR} are the ratio of atoms contributed to the dispersion at UV and IR, respectively. This equation is known as Sellmeier's equation

$$n^2 - 1 = \sum_{i=1}^{k} \frac{a_i - \lambda^2}{\lambda^2 - b_i} . \qquad (2.9)$$

Experimental data could be fitted with high precision (average error is 4.3×10^{-6}) using three term Sellmeier's equation. There are many kinds of methods for measuring refractive-index, such as minimum deviation technique, microscopic interferometry, near field pattern method and numerical appature measuring method. The minimum deviation method uses bulk glass samples and the intrinsic refractive-index can be measured without being disturbed by additional effects. Other methods, which use fibers or fiber preforms for the measurements, include the photoelastic effect because the fiber or preform has considerable stress in them.

Sellmeier's equation parameters[8,9] for pure and doped silica glass computed from the measured data of 28 wavelength points in 0.4047–2.0581 μm region are given in Table 2.4. Figure 2.15 shows the wavelength dependence of refractive-index. The refractive-index difference between pure and GeO_2-doped silica glass increases linearly with GeO_2 concentration. Refractive-index variation[10-12] for the unit molar concentration, Δn, of GeO_2 doped silica glass is shown as a function of wavelength in Fig. 2.16. The refractive-index variation increase at shorter wavelengths corresponds to the ultraviolet absorption edge shift toward longer wavelength. The remarkably small discrepancy of the absolute values of Δn is considered to be the difference in the dopant concentration estimations among different authors. Δn of P_2O_5 doped silica glass is shown in Fig. 2.17. Dispersion of Δn is very small, and this means that there are quite small absorption loss change by doping P_2O_5. Δn for B_2O_3-doped silica glass is shown in Fig. 2.18.

2.2.2 Material dispersion

Pulse spread in optical fiber is caused by material dispersion and modal dispersion. The material dispersion is given by

$$M = \frac{\lambda}{C} \frac{d^2 n}{d\lambda^2} \quad (ps \cdot nm^{-1} \cdot km^{-1}) \qquad (2.10)$$

where c is light velocity, n is refractive-index and λ is wavelength. Material dispersion can be determined using the Sellmeier's parameters in Table 2.4. Material dispersion for fused silica glass, as shown in Fig. 2.19, is 83.9 psec/nm·km and -21.9 psec/nm·km at 0.85 μm and 1.6 μm, respectively. It is zero at 1.272 μm.

Table 2.4. Sellmeier's equation parameters for pure and doped silica glass. [After Shibata[2.6]]

Sample	Dopant (mol%)		a_1	a_2	a_3	b_1	b_2	b_3	Average absolute deviation
A	—	—	0.6965325	0.4083099	0.8968766	0.004368309	0.01394999	97.93399	3.4
B	GeO_2	6.3	0.7083952	0.4203993	0.8663412	0.007290464	0.01050294	97.93428	7.8
C	GeO_2	8.7	0.7133103	0.4250904	0.8631980	0.006910297	0.01165674	97.93434	4.0
D	GeO_2	11.2	0.7186243	0.4301997	0.8543265	0.004026394	0.01632475	97.93440	7.9
E	GeO_2	15.0	0.7249180	0.4381220	0.8221368	0.007596374	0.01162396	97.93472	4.8
F	GeO_2	19.3	0.7347008	0.4461191	0.8081698	0.005847345	0.01552717	97.93484	14.0
G	B_2O_3	5.2	0.6910021	0.4022430	0.9439644	0.004981838	0.01375664	97.93353	9.3
H	P_2O_5	10.5	0.7058486	0.4176021	0.8952753	0.005202431	0.01287730	97.93401	4.3
I	GeO_2 / B_2O_3	3.4 / 9.6	0.6958807	0.4076588	0.9401093	0.004430956	0.01467544	97.93359	1.5
J	GeO_2 / P_2O_5	0.8 / 3.6	0.7026425	0.4143438	0.8952753	0.006640930	0.01094264	97.93401	5.4

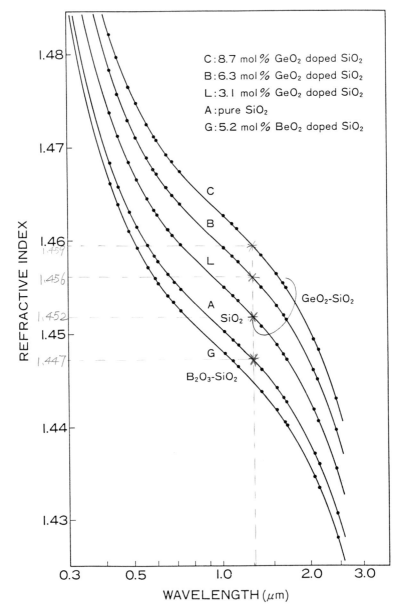

Fig. 2.15. Wavelength dependence of refractive-index for GeO_2, B_2O_3-doped and pure silica glass. [After Kobayashi[2.8]]

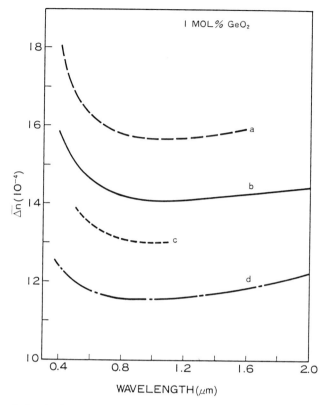

Fig. 2.16. Refractive-index variation for 1 mol% GeO_2-doped silica glass. a: Fleming[2.10], b: Shibata[2.6], c: Presby[2.11], d: Sladan[2.12]. [After Shibata[2.6]]

Material dispersion variation for unit molar concentration, material dispersion difference between pure and 1 mol% doped silica glass, is given in Fig. 2.20. The wavelength of zero material dispersion as a function of dopant concentration is given by Fig. 2.21.

2.2.3 Profile dispersion

The optimum refractive-index profile of graded-index fibers for minimizing mode dispersion depends on the wavelength dispersion properties of core and cladding glasses. When the refractive-index profile is given by

$$n^2 = \begin{cases} n_1^2\{1 - 2\Delta(\frac{r}{a})^2\} & 0 \leq r \leq a \\ n_1^2(1 - 2\Delta) = n_2^2 & a \leq r \leq b \end{cases} \quad (2.11)$$

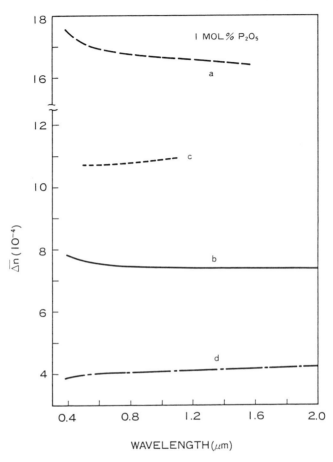

Fig. 2.17. Refractive-index variation for 1 mol% P_2O_5-doped silica glass. a: Fleming[2.10], b: Shibata[2.6], c: Presby[2.11], d: Sladan[2.12]. [After Shibata[2.6]]

The optimum value of the parameter, α_{op}, is given[13] by

$$\alpha_{op} = 2 - y - \Delta \frac{(4+y)(3+y)}{5+2y} \qquad (2.12)$$

where

$$y = -\frac{2n_1}{N_1} \cdot \frac{\lambda}{\Delta} \cdot \frac{d\Delta}{d\lambda} \qquad (2.13)$$

$$N_1 = n_1 - \lambda \, dn_1/d \, , \qquad (2.14)$$

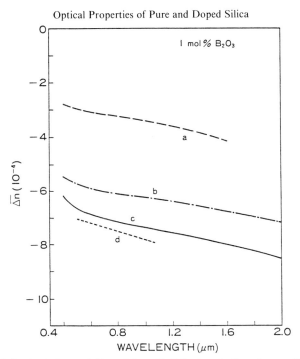

Fig. 2.18. Refractve-index variation of 1 mol% B_2O_3-doped silica glass. a: Fleming[2.10], b: Shibata[2.6], c: Presby[2.11], d: Sladan[2.12]. [After Shibata[2.6]]

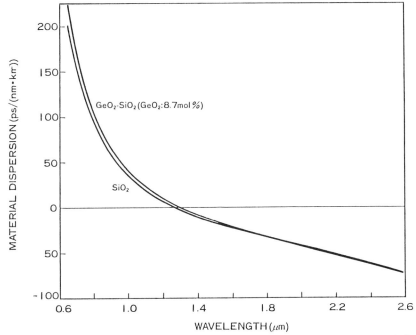

Fig. 2.19. Material dispersion for fused silica glass and GeO_2-doped silica glass. [After Shibata[2.6]]

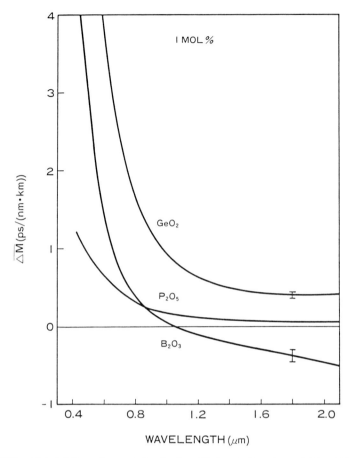

Fig. 2.20. Material dispersion variation for 1 mol% doped silica glasses. [After Shibata[2.6]]

n_1 is the refractive-index of core center, and λ is wavelength. Figure 2.22 shows the optimum profile parameter for GeO_2, P_2O_5 and B_2O_3-doped graded-index fibers.

2.2.4 Temperature dependence

Optical fiber cables are used at various temperature conditions, therefore, it is important to know the temperature dependence of the refractive-index. Refractive-index variants at high temperature[14] are shown in Fig. 2.23. Sellmeier's parameters at high temperature for pure and GeO_2-doped silica are given in Table 2.5. Refractive-index variant, $\delta n/\delta T$, of pure silica glass is

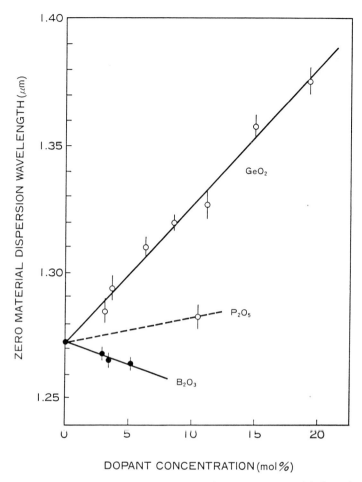

Fig. 2.21. Relationship between dopant concentration and zero material dispersion wavelength. [After Shibata[2.6]]

$1.1 \times 10^{-5}/°C$ at 22–135°C and $1.3 \times 10^{-5}/°C$ at 340–535°C at around 1.3 μm wavelength.

The temperature dependence of refractive-index dispersion is also important for the calculation of optical pulse delay time, τ, in fibers. The delay time is given by

$$\frac{\delta \tau}{\delta T} = \frac{1}{C}\left(n \cdot \frac{\delta l}{\delta T} + \frac{\delta n}{\delta T} \cdot l\right) \qquad (2.15)$$

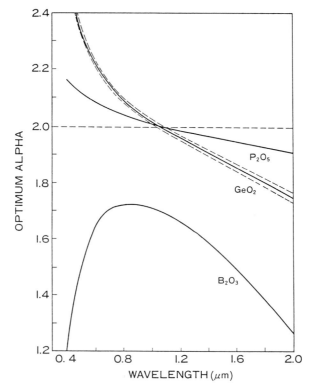

Fig. 2.22. Optimum profile parameter for GeO_2, P_2O_5 and B_2O_3-doped graded index fibers. [After Shibata[2.6]]

where l is fiber length and c is light velocity. Thermal expansion coefficient $\delta l/\delta T$ is smaller than $1\times 10^{-6}/°C$, the first term of the equation can be neglected. Material dispersion variant, $\delta\tau/\delta_T$, is 43 psec·km/°C for 15 mol% doped silica core and pure silica core fiber. Figure 2.24 shows the material dispersion of pure silica glass and 15 mol% GeO_2-doped silica glass at high temperature.

2.3 Linear Thermal Expansion Coefficients

The thermal expansion coefficients of doped silica are important parameters for designing the preform structure, because the cladding is usually made of pure silica glass, whose thermal expansion coefficient is very small; 4×10^{-7}, and that of the core glasses, doped silica glasses, are much larger. Therefore, large thermal stress will arise in the fiber preforms by the

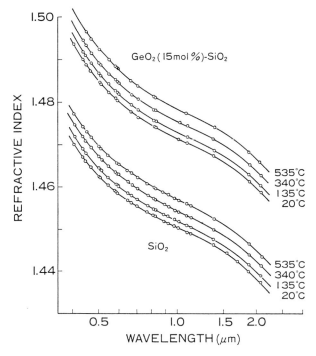

Fig. 2.23. Refractive-index variations at high temperature for SiO_2 glass and GeO_2-SiO_2 glass. [After Shibata[2.6]]

Table 2.5. Sellmeier's parameters at high temperature for pure and GeO_2-doped silica. [After Shibata[2.6]]

Sample	Suprasil W2 (20°C)	Suprasil W2 (535°C)	GeO_2(15mol%)-doped silica (20°C)	GeO_2(15mol%)-doped silica (340°C)
b_1	0.004368309	0.005922336	0.007596374	0.009087175
b_2	0.01394999	0.01253571	0.01162396	0.01029254
b_3	97.93399	97.93425	97.93472	97.93427
a_1	0.6965325	0.7053496	0.7249180	0.7301510
a_2	0.4083099	0.4174105	0.4381220	0.4437847
a_3	0.8968766	0.8705128	0.8221368	0.8042921

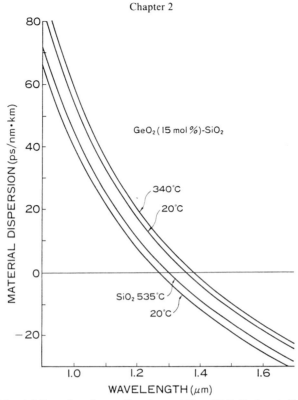

Fig. 2.24. Material dispersion of pure silica glass and 15 mol% GeO$_2$-doped silica glass at high temperature. [After Shibata[2.6]]

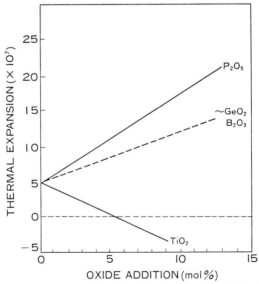

Fig. 2.25. Thermal expansion coefficient for GeO$_2$, P$_2$O$_5$, B$_2$O$_3$ and TiO$_2$-doped silica glasses. [After Izawa[2.16]]

coefficient difference. This situation is more serious in the porous preforms by OVD method, because the porous preforms are very fragile.

Figure 2.25 shows the thermal expansion coefficient for GeO_2, P_2O_5, B_2O_3, and TiO_2-doped silica glasses[16]. The coefficients increase almost linearly by increasing the dopant concentration to below 30%.

REFERENCES

1) H. Osanai, T. Shioda, T. Moriyama, A. Araki, M. Horiguchi, T. Izawa and H. Takata, "Effect of dopants on the transmission loss of low-OH-contentoptical fibers", *Electron. Lett.*, Vol. 12, p. 549 (1976).
2) T. Izawa, N. Shibata and A. Takeda, "Optical attenuation in pure and doped fused silica in the ir wavelength region", *Appl. Phys. Lett.*, Vol. 31, p. 33 (1977).
3) T. Miya, "Low-loss and low-dispersion single-mode fibers for long wavelength region" (in Japanese), Ph. D. Thesis, Tohoku Univ. (1983).
4) H. R. Phillip, "Optical transition in crystalline and fused quartz", *Solid State Commun.*, Vol. 4, p. 73 (1966).
5) P. Kaiser, A. R. Tynes, H. W. Astle, A. D. Pearson, W. G. French, R. E. Yaeger and A. H. Cherin, "Spectral losses of unclad vitreous silica and soda-lime-silicate", *J. Opt. Soc. Am.*, Vol. 63, p. 1141 (1973).
6) N. Shibata, "Optical characteristics of high silica glass fibers for communication" (in Japanese), Ph. D. Thesis, Nagoya Univ. (1983).
7) T. Izawa, "Loss-limit of high-silica fiber" (in Japanese), *1976 Joint Conv. Rec. of Four Inst. of Elec. Engineers Jpn.*, p. 3-77.
8) S. Kobayashi, N. Shibata, S. Shibata and T. Izawa, "Characteristics of optical fibers in infrared region", *Rev. ECL*, Vol. 26, p. 453 (1978).
9) N. Shibata and T. Edahiro, "Refractive-index dispersion of GeO_2, B_2O_3 and P_2O_5-doped silica glasses for optical fibers", *Trans. IECE Jpn.*, Vol. E65, p. 166 (1982).
10) J. W. Fleming, "Material dispersion in lightguide glasses", *Electron. Lett.*, Vol. 14, p. 326 (1978).
11) H. M. Presby and I. P. Kaminow, "Binary silica optical fibers: refractive index and profile dispersion measurements", *Appl. Opt.*, Vol. 15, p. 3029 (1976).
12) F. M. E. Sladen, D. N. Payne and M. J. Adams, "Profile dispersion measurements for optical fibers over the wavelength range 350 nm to 1900 nm", Tech, Dig., 4th European Conf. on Optical Fiber Commun., p. 48 (1978).
13) R. Olshanski and D. B. Keck, "Pulse broadening in graded-index optical fibers", *Appl. Opt.*, Vol. 15, p. 483 (1976).
14) N. Shibata and T. Edahiro, "Refractive-index dispersion of light-guide glasses at high temperature", *Electron. Lett.*, Vol. 17, p. 310 (1981).
15) P. C. Schultz, "Optical absorption of the transition elements in vitreous silica", *J. Am. Ceram. Soc.*, Vol. 57, p. 309 (1974).
16) T. Izawa, S. Sudo, F. Hanawa and S. Kobayashi, "Continuous fabrication process for optical fiber preforms", *ECL Tech. J.*, Vol. 26, p. 2531 (1977).

Chapter 3

FABRICATION PROCESS OF HIGH SILICA FIBERS

High silica fiber is made by drawing a preform, which has a similar structure to that of fiber to be obtained. There are various methods of preparing preforms. Modified Chemical Vapor Deposition (MCVD), Outside Vapor Deposition (OVD) and Vapor-phase Axial Deposition (VAD) methods are the most popular processes for preparing preforms. In this chapter, the details of fiber fabrication processes, mainly by the MCVD method and fiber drawing process will be mentioned.

3.1 Introduction

Optical glass fiber used in the wavelength region of 0.8–1.6 μm can be separated into two main groups from the viewpoint of the base material component: One group is high silica glasses, which consist of SiO_2 as a main component and one or more of GeO_2, P_2O_5, B_2O_3 as dopant materials. The other group is multicomponent glasses which, commonly using such optical glasses as a lens, light guide, etc., consist of one or more of the glass formers such as SiO_2, B_2O_3, or P_2O_5, modified by the incorporation of such nonglass formers as Na_2O, CaO, Al_2O_3, etc. In general, it is recognized today that the high silica glass system is better than the multicomponent glass system, especially in regard to optical transmittance, in fiber strength, and in resistance to atmosphere. Therefore, at present, high silica glass systems have occupied a prominent position as the glass systems for optical fiber use, except for special wavelength regions.

The major reason why the high silica glass systems are the most preferable materials for optical fibers is that high purity glass can be easily prepared from chloride raw materials through vapor reaction and deposition processes. The vapor phase processes share another important feature; the refractive-index control or dopant concentration control in these process is comparatively easier than that in other processes. And in these processes it is also easier to make a precise control in shape and size of fiber preforms obtained. The chemical and mechanical durabilities of the fibers are also important features for practical applications. High silica glasses have the highest durabilities in the various low-loss fiber materials.

The origin of vapor phase deposition processes can be traced to the work in the early 1930s by Hyde.[1] He invented a flame hydrolysis process for the production of articles containing vitreous silica at relatively low temperature and, if desired, of a high degree of purity. His process, as shown in Fig. 3.1, is quite similar to the Outside Vapor Deposition process[2] invented by Keck and Schultz. Fabrication process of doped silica, silica glass containing TiO_2, was invented in 1936 by Nordberg.[3] TiO_2-doped silica was originally intended as a glass having an expansion coefficient lower than that of silica. It is very interesting that TiO_2-doped silica glasses were used for the fiber core glass at the initial development stage.[2,4] The disadvantage of TiO_2–SiO_2 glass is that a very small amount of Ti^{3+} ions, which strongly absorb light in the visible and near-infrared region, is formed in the glass fabrication process. The absorption can be easily made as low as 20 dB/km by annealing the fiber in the oxygen atmosphere for few hours at 800–1000°C.[2] However, this treatment deteriorates the mechanical strength of the fiber by crystallization of silica through alkaline metal contamination.

Today, GeO_2 is used as the dopant to increase the refractive-index of silica because it seldom produces light absorbing ions in the process of glass fabrication. The one disadvantageous feature of GeO_2 is that the vapor pressure at high temperature is too high to control the concentration precisely.[5] When SiO_2–GeO_2 glass is heated to high temperature, GeO_2 evaporates from the surface. Therefore, it is very difficult to make a transparent glass directly from the raw material vapors. In order to solve the difficulty, a two-step process or soot process was developed. The process can be applied for making silica glass doped with high pressure oxides such as B_2O_3, P_2O_5, GeO_2. The soot process is "the method of making a transparent article of silica, which includes vaporizing a hydrolyzable compound of silicon into a flame of combustible gas to decompose the vapor and to form finely comminuted silica and vitrifying the silica to a transparent body".[1]

There are three major practical soot processes, Modified Chemical Vapor Deposition (MCVD),[6] Outside Vapor Deposition (OVD),[7] and Vapor-phase Axial Deposition (VAD).[8] The difference of these three processes lies in the geometrical growth direction of the fine glass particles or soot. Glass particles are deposited on a starting mandrel or tube in the radial direction in case of MCVD and OVD processes. On the other hand, glass particles are deposited on a starting rod in axial direction in VAD process. Other modification is the heat source. Although usually oxy-hydrogen gas flame are used for the chemical reaction, hydrocarbon gas flame such as CH_4–O_2, C_3H_8–O_2 or RF discharge gas plasma torch can be substituted for oxy-hydrogen gas flames.

Fabrication Process of High Silica Fibers

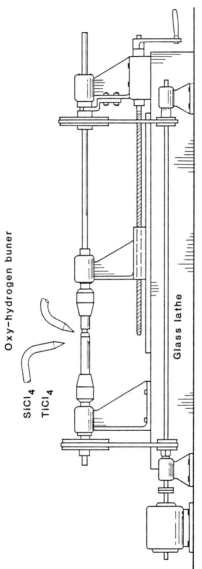

Fig. 3.1. Origin of vapor phase deposition process for glass synthesis invented by Hyde[3.1].

3.2 MCVD Process

MCVD process, developed at Bell Laboratoratories in 1974, originates from the CVD method, where the thin SiO_2 film formation process in the semiconductor industry was adopted for optical fiber fabrication:[9,11] A higher refractive-index SiO_2-GeO_2 glassy film is deposited on the inside surface of a fused silica tube by the SiH_4 and GeH_4 vapor oxidation reaction with O_2 gas. After sufficient film thickness is obtained, the tube is collapsed so that the deposited layer forms a high-index core and the tube provides the lower index cladding. The CVD method produced fibers with losses below 10 dB/km. However, this method suffered from two major disadvantages: low deposition rates and the use of hydride reactors causing OH ion contamination in the resultant glass.

Differing from the above CVD method, in 1970 Corning Glass Works presented a patent for the preparation technique (i.e., Inside Soot Deposition (ISD) method[11]) where the high-silica glass preparation technique through the soot process was applied to the fabrication of optical fibers: In this ISD method, the inside wall of a thick-walled fused silica tube is coated with a thin layer of TiO_2-doped SiO_2 core glass soot by aiming the soot stream from an oxy-hydrogen burner down through the tube hole. Then this soot layer is sintered to a bubble-free glass film. Finally, the composite is drawn into a fiber to obtain a solid glass. The first low-loss TiO_2-doped SiO_2 core fiber was made using this ISD method.[12] However, there are also several disadvantages in this method related to fiber performance, as exemplified by the requirement for 800–1000°C heating treatment to reduce the absorption loss due to Ti^{3+} ions and the difficulty in the formation of a uniform thickness soot layer along the tube.

The OVD method and MCVD method have been developed in order to overcome these problems and to produce low-OH content and high optical quality fibers.[13,14] The fundamental idea for formation of uniformly thick soot layers are the same. Fine glass particles are deposited on the surface of rotating mandrel or the inner surface of rotating silica tube. The depositing portion is moved along the axis, consequently, the uniformly thick layers are easily formed on the starting materials. Figure 3.2 shows the basic principle for making fiber by the MCVD method: A silica glass tube is rotated on a glass working lathe. Raw halide material vapors carried by oxygen gas are introduced into the silica tube, which is heated by an oxy-hydrogen flame from the outside to about 1500–1650°C. The oxy-hydrogen flame is moved repeatedly along the tube. During each traverse, the halide vapor materials are oxidized to fine glass particles and deposited on the inner surface of the silica tube. The fine glass particles deposited are immediately sintered to a transparent glass film. Multi-glass layers are formed by traversing the flame

Fabrication Process of High Silica Fibers

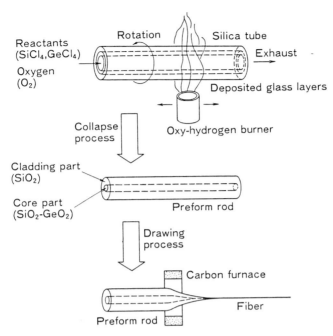

Fig. 3.2. Basic principle for making fibers using MCVD process.

repeatedly along the tube. The tube is then thermally collapsed to form a preform rod. This silica tube becomes the outer cladding and the deposited glass layers become the core and cladding of the fiber preform. By changing $GeCl_4$ concentration in the halide vapors for each traverse, the refractive-index of each glass layer can be controlled. Not only the core glass, but also the cladding glass is formed with this process. This preform is subsequently drawn into a fiber in a carbon furnace.

The MCVD method, derived from a combination of the previous CVD process and the ISD process, has several important advantages, which are related particularly to fiber performance. First, since water vapor can be completely avoided by carrying out the process in a closed space and by using halide raw materials, the resultant fiber core is essentially hydroxyl-free. Therefore it is easy to attain low-loss attenuation. Although OH diffusion from the starting silica tube increases loss in ultra-low loss fibers, increasing the barrier cladding layer thickness and decreasing the deposition temperature helps to minimize the diffusion of OH ions from the tube into the core glass.[15] Very low attenuation fibers were achieved in 1976[16] using this method. Second, by changing the halide vapor composition for each traverse of torch, it is possible to fabricate either step-index or graded-index multimode fibers. Single mode fibers with precisely controlled size and refractive-index also can be fabricated with this method. In 1979, an extremely low-loss single mode fiber was made with a loss value of 0.2 dB/km.[17] In the following sections, key points to the fabrication of high quality fibers with MCVD method will be described.

3.2.1 Instruments for MCVD process

Fundamental instruments for MCVD process are a glass-working lathe, raw material supplier and exhaust gas treatment apparatus as is shown in Fig. 3.3. The glass-working lathe is not a special one, however, the center of the two chucks should aligne coaxially so that the center difference is reduced to be less than 0.05 mm in order to prevent the silica tube deforming in the process of deposition and collapse. A driven carriage which carries oxy-hydrogen torches at a constant speed is added to the lathe. It is preferable to install a radiation thermometer and outer diameter monitor which move with the torches to measure the hot zone temperature and diameter. The heated zone can be kept at a constant temperature by controlling the flow rates of oxygen and hydrogen through the signal from the radiation thermometer.

Vapor mixtures of metal halides, such as $SiCl_4$, $GeCl_4$, $POCl_3$, BCl_3 and so on, and oxygen gas are used as the raw materials for high silica glass fabrication. Hydrogen and oxygen gases are used for heating the silica tube. The main function of the gas supply system is to precisely control the flow rates of the above vapor mixture and gases. The mixing ratio of halide vapor should be controlled to the programmed value for each layer and the combusting gases are controlled through the signal from the thermometer measuring the surface temperature of silica tube.

The metal halides used in the soot process remain in the liquid state at less than 50°C except for BCl_3, so that they are vaporized by bubbling Ar or O_2 gas through the liquid halides and are then carried into the silica tube together with oxygen. Figure 3.3(b) also shows the gas flow diagram in the supplying system. High purity halides are put into the containers made of glass or stainless steel which are respectively housed in constant temperature baths to maintain the liquid halides at the desired temperature. The flow rates of halide vapors are controlled by the flow rate of the carrier gas using thermal mass flow controllers. The flow rate of halides which are vapors at room temperature such as BCl_3, are controlled by thermal mass flow controllers without using a carrier gas.

The reacted gas exhaust from the silica tube contains fine glass particles and chlorine gas. Therefore, it is nessesary to remove these materials before exhausting to the atmosphere. Figure 3.3 shows an example of a cleaning system, which consists of soot trapper, water shower and flow control valve. Fine glass particles in the exhaust gas are first removed by the soot trapper: Next, Cl_2 and HCl are removed by the water shower along with neutralization

Fabrication Process of High Silica Fibers

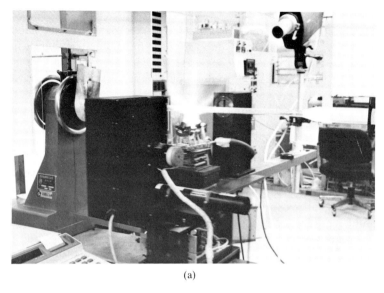

(a)

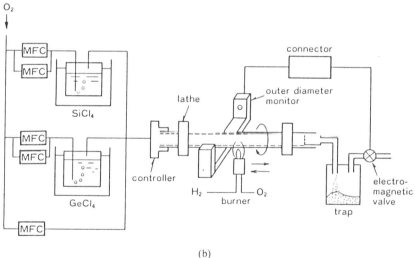

(b)

Fig. 3.3. Photograph of MCVD aparatus, (a), and schematic diagram of MCVD apparatus, (b). [After Miya[2.3]]

using NaOH. Thereby, the exhaust gas is cleaned. The exhaust gas flow is controlled by a valve, so that the pressure measured with the transducer can be kept at a constant value.

3.2.2 Tube diameter control

The fused silica tube reduces in diameter as the glass layers are deposited by the effect of surface tension. Silica tube of 14 mm outer diameter shrinks to 9 mm with about 30 deposition passes at 1600°C. The wall thickness of the tube increase with shrinking and consequently the inner surface temperature will be decreased. Figure 3.4 indicates the shrinkage of silica tube outer diameter. Sometimes, it becomes difficult to continue the deposition even if the surface temperature is high enough.

In order to avoid this trouble, it is desirable to keep the tube diameter at the initial value. The deformation of the silica tube which is mainly caused by the surface tension of the silica glass at high temperature can be prevented by applying a slight positive internal pressure. The schematic diagram of a tube diameter controller can be seen in Fig. 3.3(b). The internal pressure of the silica tube is controlled by a pressure adjusting valve near the exhaust end of the tube; the waste gas is exhausted through an orifice whose opening area is varied to control the internal pressure of the tube. The outer diameter of the silica tube in the hot zone is monitored by a noncontacting-measuring set utilizing a deflecting He–Ne laser beam. The signal is fed back to the pressure adjusting apparatus in order to control the outer diameter to the desired value. Figure 3.5 exhibits the marked effect of this tube diameter control system.

3.2.3 Prebake of silica tube

There are many surface imperfections such as small scratches and bubbles on the inner surface of the tube as shown in photograph of Fig. 3.6,

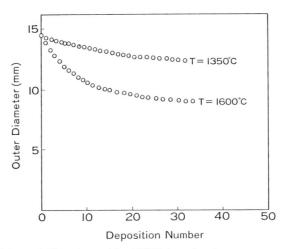

Fig. 3.4. Shrinkage of silica tube used for MCVD heated at the temperatures of 1350°C and 1600°C. [After Miya[2,3]]

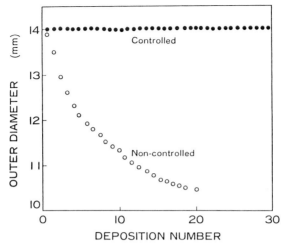

Fig. 3.5. Effect of tube diameter control system. [After Miya[2.3]]

Fig. 3.6. Photograph of small scratches and bubbles on the inner surface of silica tube. [After Miya[2.3]]

which can be observed as a scattering center when a part of tube is heated strongly. They affect the scattering loss of the produced fibers. Typical additional loss caused by these surface imperfections is estimated to be 0.5–1.5 dB/km. The experimental relation between the imperfection density and scattering loss is shown in Fig. 3.7

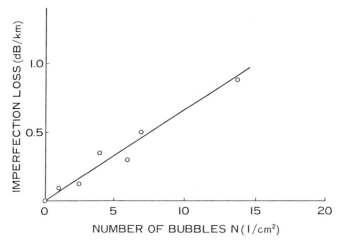

Fig. 3.7. Relationship between bubble density of tube surface and imperfection loss of optical fiber. [After Miya[2.3]]

The scratches and bubbles can be removed by prebaking the tube at high temperature, previous to the deposition. Figure 3.8 shows the decrease of the scattering centers by prebaking. The optimum temperature is about 1600°C. At higher temperatures, scattering center density decreases abruptly, however, a small number of centers remain. It appears that some of the

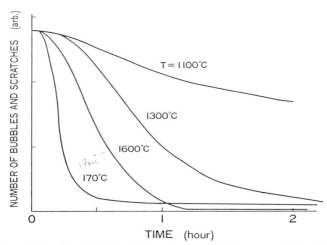

Fig. 3.8. Reduction of bubbles in silica tube by prebaking process. [After Miya[2.3]]

scratches create new bubbles because the scratches are covered with a glass layer before it is removed by surface tension.

3.2.4 Hydroxyl impurity reduction

The hydroxyl contamination in the glass fibers may be attributed to two causes. One is the contamination due to hydrous and hydroxyl impurities in the raw material chlorides, and water vapor in the carrier gas. The other is OH diffusion from the silica tube.

Major hydrous impurity in $SiCl_4$ is trichlorosilane ($SiHCl_3$). Trichlorosilane can be removed by multistep distilation. The OH absorption loss at 1.39 μm in a fiber made from $SiCl_4$ is shown in Fig. 3.9, in terms of the relation of the partition fractions for $SiHCl_3$ within the raw materials.[18,19] The losses are proportional to the $SiHCl_3$ concentration in the raw materials, but it does not go through the origin. The background level of OH loss is due to other impurities, such as HCl, CH_4 and H_2O. Water vapor present in the carrier gas for the chlorides, oxygen or argon, is also an important cause of OH contamination. To remove water vapor in the gas, it is very important not only to purify the gas itself, but also to prevent leakage of air at tube connection. The rotating connector used at the raw material inlet of silica tube and the plastic fittings are the major cause of leaks.

OH ions in the silica tube diffuse into deposited silica glass during the process of deposition. The diffusion length is about 47 μm before collapsing. This is in fairly good agreement with the expected value of 34 μm, calculated

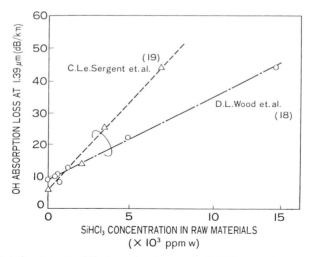

Fig. 3.9. Relation between OH-absorption loss peak at 1.39 μm and amount of $SiHCl_3$ contained within the raw materials. The data are adduced from Refs. (3.18), (3.19).

by taking a diffusion constant of 7.3×10^{-9} cm^{2}sec^{-1} at 1600°C and $t=400$ sec, and integrating the heating time during the fabrication.[20] Figure 3.10 shows OH ion distribution profiles, evaluated from the optical density at 2.73 μm, in four preforms having different thickness of deposited silica glass of (a) 1.0 mm, (b) 1.5 mm, (c) 2.0 mm and (d) 2.5 mm. The distance from the inner surface of deposited tube before collapsing is also indicated at the upper side of the figure.

It is clear from this data that the buffer layers are needed to prevent OH contamination from silica tube. OH absorption loss of single mode fibers changes with the thickness of deposited cladding layers because a fair proportion of the light power propagates in the cladding. Figure 3.11 shows the OH absorption at 1.39 μm of single mode fibers with different cladding thickness. It is clear that the ratio of cladding diameter to core diameter should be larger than 5 to obtain a low-loss fiber.

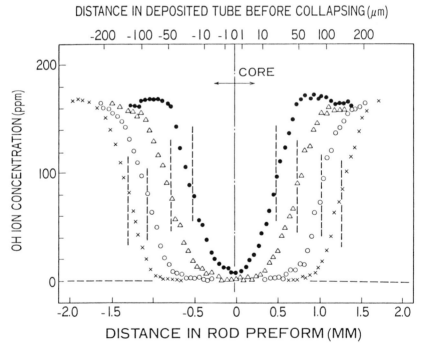

Fig. 3.10. OH-ion distribution profiles in four MCVD preforms having different thickness of deposited silica glass; (a) 1.0 mm; (b) 1.5 mm; (c) 2.0 mm; (d) 2.5 mm. [After M. Kawachi, M. Horiguchi, A. Kawana and T. Miyashita, OH-ion distribution profiles in rod preform of high silica optical waveguides, *Electron. Lett.*, Vol. 13, p. 247 (1977)]

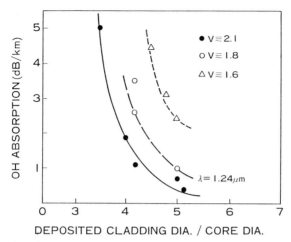

Fig. 3.11. Relationship between OH-absorption peak at 1.39 μm and the ratio of deposited cladding diameter to core diameter. [After Miya[2.3]]

3.2.5 Optimum deposition temperature

The transmission loss of MCVD fibers is strongly affected by the preform fabrication conditions, mainly glass deposition temperature. When the temperature is too low, the glass particles deposited can not be sintered to a transparent glass, and when it is too high, the silica tube shrinks too much to continue deposition. The temperature is divided into the following four regions:

(1) Appropriate Temperature T_{ap}
(2) Intermediate Temperature T_{im}
(3) Critical Temperature of Bubble Creation T_{cr}
(4) Bubble Creation Temperature T_b

T_b is the temperature where the scattering center in the glass layers can be seen easily in the deposition process. T_{cr} is the temperature where the scattering center is barely seen. T_{ap} is the temperature where the scattering center is perfectly eliminated as shown in Fig. 3.12. Just above T_{cr}, scattering centers still remain in the glass layers and the structural scattering loss is clearly recognized in the loss characteristics.[21] The microscopic surface of the glasses made at the four different temperature are quite different. Figure 3.13 shows the electron microscope pictures of the glass surface made at the temperature of T_{ap} (1650°C), T_{im} (1615°C), T_{cr} (1580°C) and T_b (1500°C), respectively. These temperatures vary slightly with the glass composition, it is very important to deposit glass layers at an optimum temperature in order to fabricate low-loss fibers.

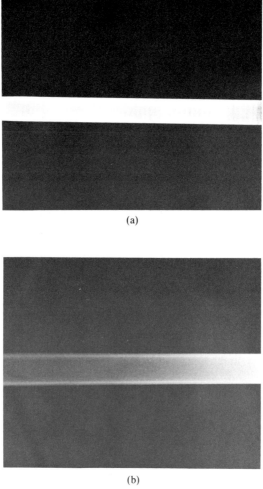

Fig. 3.12. Photographs of silica tube inner surface in the deposition process; (a) is at bubble creation temperature and (b) is at the appropriate temperature. [After Miya[2.3]]

3.3 Outside Vapor Deposition Process

Outside Vapor Deposition process was the first successful fabrication process of high silica fibers. The process is shown schematically in Fig. 3.14. A vapor mixture of chloride raw materials is reacted in a oxy-hydrogen flame to produce fine glass particles of the desired composition. The flame is directed toward a ceramic target rod, which is rotating and traversing in a lathe. The

Fabrication Process of High Silica Fibers 65

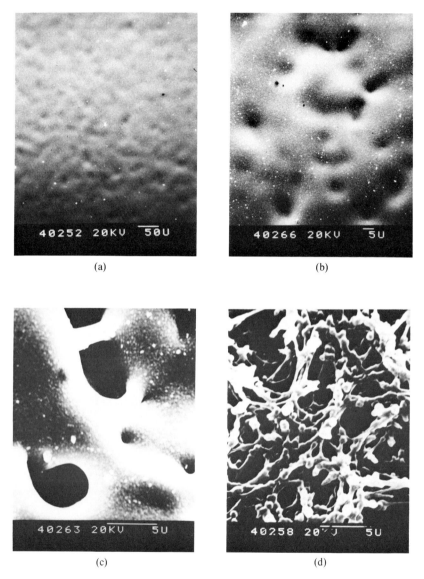

(a)

(b)

(c)

(d)

Fig. 3.13. Electron microscope pictures of the glass surfaces made at the temperatures of (a) T_{ap} (1650°C), (b) T_{im} (1615°C), (c) T_{cr} (1680°C), (d) T_b (1500°C). [After Miya$^{(2.3)}$]

1580°C

glass particles stick to the rod in a porous layer. The composition of the glass particles is controlled layer by layer in order to form the radial refractive-

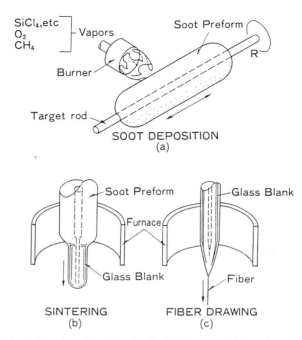

Fig. 3.14. Principal step in making fibers by the OVD process. (a) Soot deposition process. (b) Preform sintering process. (c) Fiber drawing from the glass blank. [After P. C. Schultz, Fabrication of optical waveguide by the outside deposition process, *Proc. IEEE*, Vol. 68, p. 1181 (1980) Copyright © 1980 IEEE]

index profile. Both step-index and graded-index fibers can be readily achieved using this approach.

When the deposition is completed, the porous preform is slipped off the ceramic target rod and then sintered to a transparent bubble-free preform in a dry inert atmosphere at about 1500°C.

Although the details of the OVD process have not been reported, it is estimated from the other processes that the removal of the target rod and cracking of the porous preform by thermal stress are major technical problems. Other processes such as consolidation and dehydration are almost the same as in the VAD process described in the next chapter.

3.4 Fiber Drawing

Optical fibers are made simply by drawing heated preforms. One of the most important things is to maintain the mechanical strength of silica glass. The mechanical strength of bulk silica glass is essentially as large as 7 GPa, the value is twice as large as that of piano wire. The major origin of reducing optical fiber strength is thought to be the presense of surface flaws. Most of the surface flaws are formed in the fiber drawing process. In this section, some of the practical processes for making high-strength fibers are described.

3.4.1 Drawing apparatus

Figure 3.15 shows the schematic diagram of the drawing apparatus, which is composed of preform carrier, electric furnace, outer diameter measurement instrument, plastic coating applicator, plastic curing furnace, capstan and take-up drum. The preform carrier is to feed the preform into the electric furnace at a constant speed. The furnace is heated above 2000°C to soften the preform end. The heater is usually made of high purity carbon or high purity zirconia. Figure 3.16 shows a schematic diagram of carbon resistance furnace. The temperature fluctuation in the furnace directly causes the fiber diameter to vary. Therefore, the temperature must be controlled precisely. Furthermore, the dust in the furnace causes surface flaws, so the

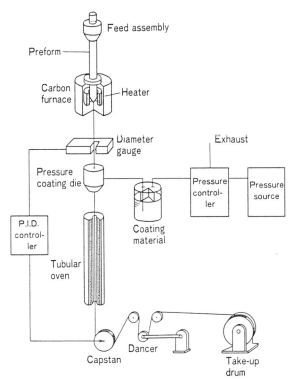

Fig. 3.15. Layout diagram for fiber drawing and coating system. [After Chida[3.22]]

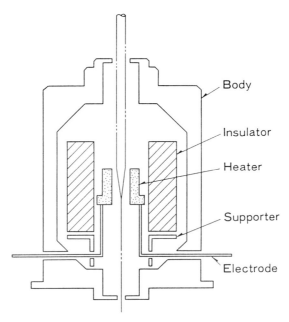

Fig. 3.16. Schematic diagram of carbon resistance furnace for fiber drawing. [After Sakaguchi[3.23]]

atmosphere in the furnace should be clean. The plastic coating applicator and the curing furnace are used for forming the primary coating layer to prevent the additional formation of surface flaws by touching the fiber surface on the capstan and take-up drum. The take-up speed of the fiber is controlled by the signal from the outer diameter measurement apparatus so as to maintain the outer diameter at a constant value.

3.4.2 Diameter control

There are two major origins of fiber diameter fluctuation, one is short term fluctuation caused by temperature fluctuation of the furnace and the other is long term fluctuation cased by the preform outer diameter fluctuation. The short term fluctuation, whose periodic length is less than about 100 cm, can be reduced to less than ± 0.5 μm by decreasing the temperature fluctuation to less than $\pm 0.2°$C. In order to realize precise temperature control, inert gas flow rate in the furnace and the ratio of preform diameter to heater inner diameter must be optimized. The long term fluctuation, whose periodic length is longer than 100 cm, can be reduced by using a feed-back control to the taking-up speed through the signal from the outer diameter measurement apparatus. Figure 3.17 shows typical examples of fiber diameter.

3.4.3 Strength

The strength of drawn fiber changes with the surface treatment of preforms, dust particle density in the furnace, humidity of the furnace atmosphere and primary coating thickness.

A) Surface treatment

The surface strain of the preform cause the reduction of tensile strength. In order to highlight the surface treatment effectiveness, silica fibers were drawn from fused silica rods treated before drawing with various kinds of surface treatment methods. These methods include:

A; no treatment (only rinsed with acetone).
B; flame-polished (flame-polished with an oxy-hydrogen flame at 1900°C).
C; HF treated and flame-polished (immersion in 49 wt% hydrofluoric acid, rinsed with deionized distilled water and then flame-polished with an oxy-hydrogen flame).

Figure 3.18 shows Weibull distribution curves indicating tensil strength for silica fibers drawn with surface treated silica rods. In case of treatment A, the strength is widely distributed from weak to strong regions. In case of using

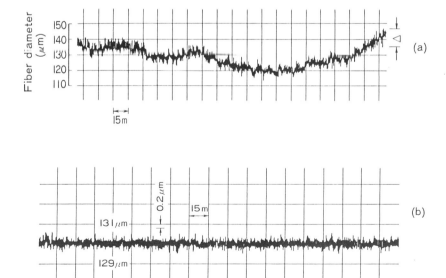

Fig. 3.17. Typical examples of fiber diameter fluctuation; (a) uncontrolled; (b) precisely controlled. [After Sakaguchi[3.23]]

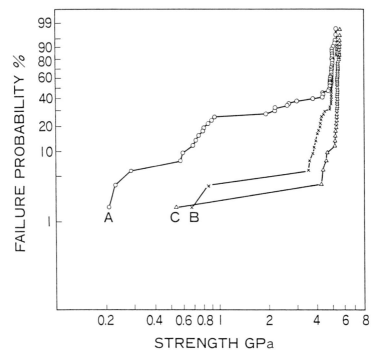

Fig. 3.18. Strength distribution of fibers drawn from surface treated preforms: A; no treatment, B; fire polished, C; HF treated and fire polished. [After Sakaguchi[3.23]]

silica rods immersed in hydrofluoric acid and flame-polished through oxy-hydrogen flame, the fiber is strong and the obtained strength distribution is almost unimodal. These investigations indicate that preform surface treatment is effective to obtain high-strength fibers.

B) *Dust particles in the furnace*

Fine particles are released from the refractory materials such as carbon, zirconia, and alumina. Although no dust particles are generated when no electric power is supplied, dust particles are generated and the density increases rapidly when power is supplied. A typical density is 1–10 particles/liter, and the size is 0.3–1 μm. These particles cause a reduction of the tensile strength of the fibers.

An experimental relation between dust particle size in the furnace and the tensile strength of the drawn fiber is that the tensile strength is proportional to the reciprocal root of dust particle radius. A simple solution for dust reduction is to flow filtered gas into the furnace and to close the furnace by using a clean gas shutter. Figure 3.19 shows the gas flow effect.

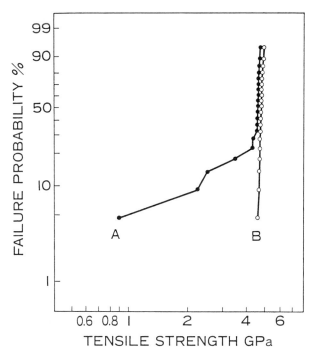

Fig. 3.19. Effect of gas flow port on strength, clean gas flow prevents the surface contamination. [After Sakaguchi[3.23]]

C) *Humidity*

The humidity of the furnace atmosphere also affects the drawn fiber strength. Figure 3.20 shows Weibull distribution curves indicating tensile strength for fibers drawn under two different conditions; one group is drawn with flowing dry argon, and the other is drawn with flowing argon saturated with water vapor. Clear difference in the strength was observed in these experiments. We can say that it is necessary to maintain a flow of dried gas in the furnace for making high-strength fibers.

D) *Primary coating*

Plastic thin layer should be coated on the drawn fiber just after drawing and before touching the capstan and other parts of drawing machine in order to prevent the formation of flaws on the fiber surface. The thickness of the primary coating is of fundamental importance. Figure 3.21 shows the thickness dependence of the tensile strength. Usually, a 100 μm thick silicone layer is sufficient for this purpose.

In the fiber drawing process, the drawing speed is usually limited by the coating speed. A plastic thin layer is coated on a fiber by dipping in silicon

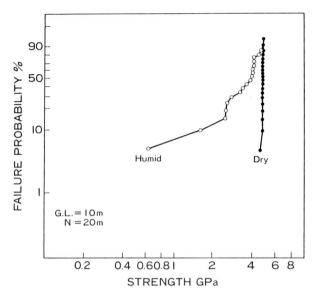

Fig. 3.20. Effect of water vapor in the drawing furnace on fiber strength. Humid atmosphere degrades the fiber surface. [After Sakaguchi[3.23]]

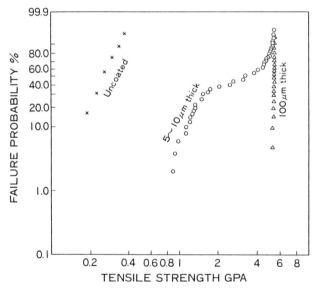

Fig. 3.21. Effect of primary coating thickness on fiber strength. Very thin film does not work for the surface protection during drawing process. [After Sakaguchi[3.23]]

resin in an open die. When the coating speed is low, the fiber is smoothly coated. On the other hand, when the coating speed exceeds a critical speed, the fiber slips and the plastic does not coat. Figure 3.22(a) shows the silicone resin movement observed when the fiber is smoothly coated. At high drawing speed, the resin slips and the convection flow can not be observed as is shown in Fig. 3.22(b). For high speed coating application, a pressurized die as shown in Fig. 3.23 has been developed. The coating material, contained in a high

(a)   (b)

Fig. 3.22. Photographs visualizing silicone resin movement in applicator: (a) drawing speed 60 m/min, (b) drawing speed 120 m/min. In case of (a), air bubbles are introduced on the fiber surface. [After Chida[3.22]]

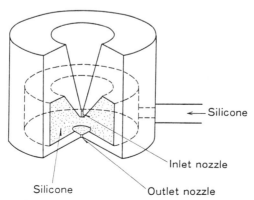

Fig. 3.23. Schematic diagram of pressurized die for primary coating. [After Chida[3.22]]

pressure container, was first pumped by an external high pressure gas source. Then, the coating material was fed into the die. High speed coating at 360 m/min has been achieved with this apparatus.[22]

REFERENCES

1) J. F. Hyde, "Method of making a transparent article of silica", U.S. Patent 2,272,342 (Filed August 27, 1934).
2) D. B. Keck and P. C. Schultz, "Method of forming optical waveguide fibers", U.S. Patent 3,737,292 (Filed Jan. 3, 1972).
3) M. E. Nordberg, "Glass having an expansion lower than that of silica", U.S. Patent 2,326,059 (Filed Apr. 22, 1936).
4) F. P. Kapron, D. B. Keck and R. D. Maurer, "Radiation losses in glass optical waveguides", *Proc. of Conference on Trunk Telecommunicaitons by Guided Wave*, Sep. 29–Oct. 2, 1970, pp. 148–153.
5) G. V. Samsov, *The Oxide Handbook* (IFI/Plenum, New York-Washington-London, 1973), p. 178.
6) J. B. MacChesney, "Materials and progresses for preform fabrication—Modified Chemical Vapor Phase Deposition and Plasma Chemical Vapor Deposition", *Proc. IEEE*, Vol. 68, No. 10, pp. 1181–1184 (1980).
7) P. C. Schultz, "Fabrication of optical waveguide by the outside vapor deposition process", *ibid.*, pp. 1187–1190.
8) T. Izawa, S. Sudo and F. Hanawa, "Continuous fabrication process for high silica fiber preforms", *Trans. IECE Jpn.*, Vol. E62, No. 11, pp. 779–785 (1979).
9) J. B. MacChesney, P. B. O'connor, J. R. Simpson and F. V. DiMarcello, "Multimode optical waveguide having a vapor deposition core of germania doped bolosilicate glass", *Am. Ceram. Soc. Bull.*, Vol. 52, p. 704 (1973).
10) J. B. MacChesney, R. E. Jaeger, D. A. Pinnow, F. W. Ostermayer, T. C. Rich and L. G. Van Uitert, "Low-loss silica-bolosilicate clad fiber optical waveguide", *Appl. Phys. Lett.*, Vol. 23, No. 6, pp. 340–341 (1973).
11) D. B. Keck and P. C. Schultz, "Method of producing optical waveguide fibers", U.S. Patent 3,711,262 (Filed May 11, 1970).
12) P. C. Schultz, "Vapor phase materials and processes for glass optical waveguides" in *"Fiber Optics"*, B. Bendow and S. S. Mitra (ed.), Plenum Press, New York and London (1979).
13) L. B. MacChesney, "Preparation of low loss optical fibers using simultaneous vapor phase deposition and fusion", *Proc. 10th Int. Congr. Glass*, Vol. 6, pp. 40–44 (1974).
14) W. G. French, R. E. Jaeger, J. B. MacChesney, S. R. Nagel, K. Nassau and A. D. Pearson, "Fiber preform preperation", in *"Optical Fiber Telecommunication"*, S. E. Miller and A. G. Chynoweth (ed.), Academic Press, New York, San Francisco, London (1979).
15) M. Horiguchi and M. Kawachi, "Measurement technique of OH-ion distribution profile in rod preform of silica based optical fiber waveguides", *Appl. Opt.*, Vol. 17, No. 16, pp. 2570–2574 (1978).
16) M. Horiguchi and H. Osanai, "Spectral losses of low-OH content optical fibers", *Electron. Lett.*, Vol. 12, pp. 310–312 (1976).
17) T. Miya, Y. Terunuma, T. Hosaka and T. Miyashita, "Ultimate low-loss single-mode fiber at 1.55 μm", *Electron. Lett.*, Vol. 15, pp. 106–108 (1979).
18) D. L. Wood, T. Y. Kometani, J. P. Luongo and M. A. Saifi, "Incorporation of OH in glass in the MCVD process", *J. Am. Ceram. Soc.*, Vol. 62, No. 11–12, pp. 638–639 (1979).
19) C. Le. Sergent, M. Liegois and Y. Floury, "Influence of purity of reactants on the attenuation of MCVD made doped silica optical fibers", *J. Non-Cryst. Solids*, Vol. 39, pp. 263–268 (1980).

20) M. Kawachi, M. Horiguchi A. Kawana and T. Miyashita, "OH-ion distribution profiles in rod preform of high silica optical waveguides", *Electron. Lett.*, Vol. 13, p. 247 (1977).
21) T. Miya, Y. Terunuma, T. Hosaka and T. Miyashita, "Fabrication technique of single-mode optical fibers" (in Japanese), *ECL Tech. J.*, Vol. 28, No. 6, p. 945 (1979).
22) K. Chida, T. Kimura and M. Wagatsuma, "High-speed optical fiber drawing", *Rev. ECL*, Vol. 32, p. 425 (1984).
23) S. Sakaguchi, "Studies on strength characteristics for high-silica optical fibers", Ph. D. Thesis, Tohoku Univ. (1984).
24) T. Miya, "Studies on loss reduction and dispersion minimization of single mode optical fibers for long-wavelength region", *ibid.* (1983).

Chapter 4

VAPOR-PHASE AXIAL DEPOSITION PROCESS

4.1 Introduction

The Vapor-phase Axial Deposition (VAD) process has been developed by combining the doping technique of the OVD process and the axial deposition technique in a direct deposition process, yielding the conceptional basis of fabrication process simplification and fiber preform mass production.[1-5] The key feature in the VAD method is the axial growth of fiber preforms. Hence, this process, in principle, allows continuous optical fiber preform fabrication, which seems to be impossible in the conventional processes. Figure 4.1 shows the basic principle for making fiber preforms by

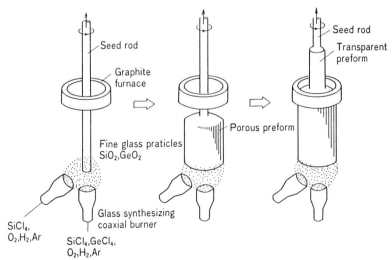

Fig. 4.1. Schematic diagram of VAD process. [After Izawa et al.[4.5]]

the VAD method: Raw halide materials, blown from a torch, oxidize to fine glass particles or soot by the flame hydrolysis. Both core glass and cladding glass soot are simultaneously deposited in an axial direction onto the end of a rotating fused silica target rod. As the rod-like porous preform grows devoid of a central hole, differing from the OVD method, the preform is slowly retracted through a graphite resistance furnace, where it is consolidated to a transparent preform by zone sintering.

The adaption of VAD process mentioned above provides the feasibility of several important advantage related to both mass production and fiber performance. Since the VAD process has essentially the same soot process as the OVD process, the effective glass deposition rate is high. There is a potential for mass production, because of both fabrication process simplicity and the capability of continuous preform fabrication. Further, if the fabrication condition can be successfully kept constant throughout the preform preparation, the preform has a good uniformity in glass material and hence an excellent attenuation reproducibility. In addition, VAD fibers have no dip in the center of the refractive index profile, which is commonly observed in MCVD fibers and OVD fibers that deteriorate bending loss characteristics in single mode fibers.

On the other hand, technical problems foreseen in the VAD process are as follows. First, it seems to be difficult to prepare the rod-like porous preform, made only of fine glass particles without a mandrel. Second, it appears to be difficult to completely consolidate porous preforms and effectively reduce OH-ion impurity in the preform resulting from the flame hydrolysis products. Third, it appears to be difficult to accurately control the glass composition gradient (and thus, refractive-index profile) across the blank core and hence to obtain wide-bandwidth graded-index fibers, unlike the OVD and MCVD processes.

Most of these problems have been solved and the performance of fibers made by the VAD process is similar to fibers made by the MCVD and OVD processes. Furthermore, nowadays top performance of transmission loss and bandwidth have been reported with fibers fabricated by the VAD process. In this chapter, the details of the VAD process will be described.

4.2 Preform Fabrication Apparatus

The VAD preform fabrication apparatus construction is shown in Fig. 4.2. As shown, the VAD preform fabrication apparatus consists of the following components:[4,5] 1) A system for supplying raw materials and flame combustion gases, 2) porous preform preparation chamber, 3) electric furnace for consolidation, 4) pulling mechanism and its positional control system and 5) exhaust system. Raw halide materials, such as $SiCl_4$, $GeCl_4$, $POCl_3$, etc., are carried by Ar gas from the raw halide material supply system and fine glass

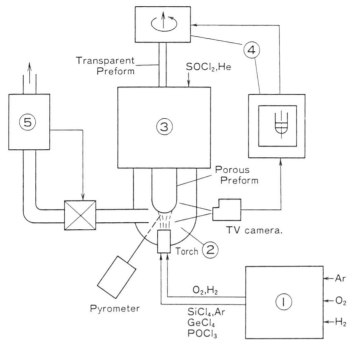

Fig. 4.2. VAD preform fabrication apparatus construction. (1) Supply system for raw materials and flame forming gases. (2) Porous preform preparation system. (3) Electric furnace for consolidation. (4) Pulling mechanism and positional control system. (5) Gas exhaust system. [After Sudo[4.57]]

particles are synthesized by the flame hydrolysis reaction of these raw halide vapors blown from the torch. The rod-like porous preform is grown by the deposition of fine glass particles onto the end of the target silica rod in the direction of its rotating axis.

The growing end surface temperature of this porous preform formed is measured with the pyrometer to achieve the index-profile control in the core preform as detailed in Section 4.6. The porous preform is continuously consolidated into the transparent preform by zone sintering in the carbon-ring electric furnace. The transparent preform is pulled upward by the pulling mechanism rotating around its axis, while the pulling speed is controlled so as to keep the porous preform growing end at a fixed position, measured with the TV-monitoring system. In addition, the undeposited glass particles, and H_2O and HCl gases caused by the flame hydrolysis are removed by the exhaust system to maintain the uniformity in the porous preform diameter. Figure 4.3 is a photograph of a VAD apparatus.

Fig. 4.3. Photograph of VAD apparatus. [After Sudo[4.57]]

4.2.1 Porous preform fabrication chamber

The porous preform fabrication chamber contains torches and pyrometer. In this section, the structure of torches and chamber are described.

A) *Torches*

The torches for synthesizing fine glass particles are usually made of silica glass and basically have four concentric nozzles, as shown in Fig. 4.4:[6,7] The halide material vapor mixture flows from nozzle I. Inert gas, such as argon, flows from nozzle II. Hydrogen gas for flame combustion flows from nozzle III. Oxygen flows from nozzle IV. A typical example of the gas supply condition for graded-index fiber fabrication is given below:[6]

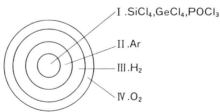

Fig. 4.4. Cross sectional view of fundamental torch structure. [After Sudo[4.57]]

Halide materials from nozzle I: Ar gas at a flow of 100 cc/min saturated with 40°C $SiCl_4$ and Ar gas at a flow of 100 cc/min saturated with 30°C $GeCl_4$. Inert gas from nozzle II: 1000 cc/min of Ar gas. Combustion gas from nozzle III: 3000 cc/min of H_2 gas. Auxiliary gas from nozzle IV: 4000 cc/min of O_2 gas. The main reason for using this torch is to blow the inert gas from nozzle II, which has marked effects on adjusting the flame temperature as well as preventing fine glass particles deposition on the torch end.

Several torch structures were invented, based on this fundamental structure, and examples are shown in Fig. 4.5. Torches (a) and (b) were devised for controlling the refractive-index profile in the early stage of the VAD development:[7,8] Different composition vapor mixtures flow from each nozzle, indicated by reference numeral I, and are mixed with each other. As a result, the dopant concentration distribution is formed spatially. Torch structure (c) was designed to fabricate single-mode fiber preforms.[9] The key point of this torch is that it has a nozzle for blowing halide materials set at a position deviated from the center of the torch flame. Thereby, the core diameter is sufficiently reduced to enable the deposition of a thick cladding layer on the periphery of the core porous rod.

B) Reaction chamber

The reaction chamber is usually made of glass with a cubic-shaped bottom. It has two windows for observation and for measuring the surface temperature and position of the growing end of the porous preform, and a gas exhaust port. The important points for use of the reaction chamber are to keep the inside environment clean, and to keep the gas flow unperturbed.

A pyrometer is used to measure the preform surface temperature. Two dimentional measurement of the surface temperature is preferable, because, in the VAD process for GeO_2-doped fibers, the index-profile is mainly formed by the two-dimensional temperature effect on the porous preform growing surface. Scanning type pyrometers are good for this purpose.

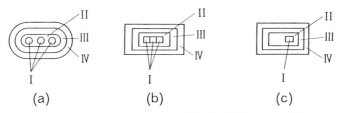

Fig. 4.5. Various kinds of torches for soot deposition. I; Halide material blowing nozzle, II; Inert gas blowing nozzle, III; Combustible gas blowing nozzle, IV; Auxiliary gas blowing nozzle. [After Sudo[4.57]]

4.2.2 Consolidation furnace

The porous preform prepared in the chamber is usually consolidated into a transparent preform by a zone sintering technique. A cross sectional view of the zone-sintering consolidation furnace is shown in Fig. 4.6. The heater is made of high-purity carbon, driven by a DC power source. Heat insulating material is made of high-purity carbon pipe. The vessel is made of stainless-steel. Ar gas is introduced into the vessel to provide an inert atmosphere. He gas, important to promote porous preform consolidation, and $SOCl_2$ gas or Cl_2 gas for dehydration treatment, are also introduced into the furnace tube. Figure 4.7 indicates a temperature distribution for this consolidation furnace. Maximum temperature is typically about 1500–1600°C and the temperature gradient along its axis is about 140°C/cm around the center part of the carbon ring heater. This temperature gradient and gas-tightness of the furnace tube play an important role in the achievement of the complete consolidation and

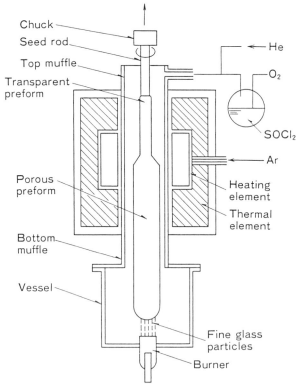

Fig. 4.6. Cross sectional view of zone-sintering consolidation furnace. [After K. Chida, Simultaneous dehydration with consolidation for VAD method, *Electron. Lett.*, Vol. 15, No. 25, p. 835 (1979)]

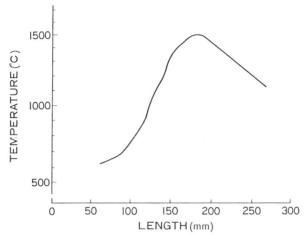

Fig. 4.7. Temperature profile along the axis of consolidation furnace. [After Sudo[4.57]]

OH-ion reduction as detailed in Section 4.5. The consolidation can be also carried out using an electric furnace having a uniform temperature distribution.

4.2.3 Pulling mechanism

A preform pulling machine is the basis of the VAD preform fabrication apparatus. This pulling machine, requires both structural stability and dimensional accuracy at as high a level as that for a single crystal pulling machine. In order to fabricate high-quality preforms continuously, it is essential to realize i) uniformity in the target rod rotating speed, ii) straightness in the pulling axis and iii) precise control in the pulling speed. Figure 4.8 shows the construction of a VAD pulling machine.[5] In this machine, the target rod positional fluctuation along with the rotation, which has an influence on the cylindrical symmetry of the preform, is within ±0.05 mm or less at the particle deposition position. The positional variation of the rotating axis which has a significant affect on the uniformity of the preform lengthwise, is reduced to within ±0.1 mm for 1 meter axial length. The porous preform growing end, where the positional fluctuation causes preform diameter variation and index profile fluctuation, is kept at a fixed position typically within a 50 μm accuracy by a feedback mechanism, using a positional signal attained from a video camera.

There are two other important factors in the pulling mechanism related to continuous preform fabrication. One is concerned with the pulling speed for continuous consolidation and the other is concerned with the continuous

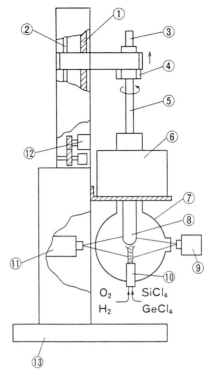

Fig. 4.8. Schematic diagram of preform pulling machine. (1) Ball screw for up-down driving, (2) Slade shaft, (3) Motor for seed rod rotation, (4) Chuck for seed rod, (5) Seed rod, (6) Consolidation furnace, (7) Glass vessel, (8) Porous preform, (9) Temperature monitor, (10) Torch, (11) TV camera for position monitoring, (12) Motor for preform pulling, (13) Base. [After Sudo[4.57]]

pulling technique. Figure 4.9 is a diagram showing continuous consolidation in the VAD process.[11,12] The pulling speed, V, for continuous consolidation is expressed as

$$V = a(1 + b)V_0 \qquad (4.1)$$

where V_0 is the porous preform growing speed, a is the porous preform shrinkage rate in the axial direction and b is the transparent preform elongation rate at the neck down region due to the porous preform weight. It is found, from eq. (4.1), that the pulling speed, V, is closely connected with both the porous preform fabrication conditions and the consolidation conditions, and is variable during continuous fabrication. Therefore, the above-

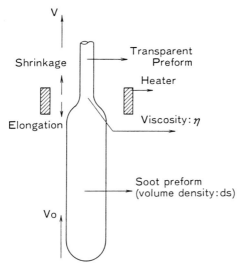

Fig. 4.9. Continuous consolidation in the VAD method. [After Nakahara[4,12]]

mentioned pulling speed feedback control system has a marked effect on the production of uniform preform diameter and index profile.

4.2.4 Exhaust system

The main exhaust system functions are: i) Removal of gases, such as H_2O and HCl caused by the flame hydrolysis reaction, ii) cleaning the resultant exaust gas and iii) flow rate control of the exhaust gas.[13]

Fine glass particles in the exhaust gas are first removed off by the soot traps: Next, HCl and H_2O are removed by the water shower along with neutralization of the HCl using NaOH. Thereby, cleaning the exhaust gas. The exhaust gas flow rate is controlled by a valve, so that the pressure in the chamber can be kept at a constant value. This exhaust system also plays an important role in achieving the porous preform uniformity, as described in Section 4.3.

4.3 Porous Preform Fabrication

Techniques to prepare porous preforms in an axial direction are not only one of the most fundamental keys for fiber preform fabrication by the VAD method, but also of extreme importance in relation to the transmission characteristics of the resultant optical fibers and to the axial uniformity in the preform diameter. There seems, however, to be a great deal of technical difficulty in forming the rod-like porous body composed of only fine glass particles in the axial direction. To put this in more precise terms, it is only in the early stage of porous preform preparation that the target rod has an influence on the preform dimensions and on the growing surface shape, while, in the steady state, the preform dimension and the growing front shape might be determined as a result of the complicated interaction among various preparation factors, such as the fine particle stream, the growing surface temperature and the supply rate of starting materials.

The VAD porous preform preparation process comprises of the following four main processes: 1) Synthesis of fine glass particles, 2) particle deposition, 3) porous preform preparation in an axial direction, and 4) size control. In this section, porous preform preparation techniques are described in detail, based on the above four processes. Both the simultaneous formation technique for the porous cladding and core, and the preparation technique for porous preform for single mode fiber fabrication are also presented.

4.3.1 Synthesis and deposition of fine glass particles

Oxide products (such as SiO_2, GeO_2, P_2O_5) are synthesized by the flame hydrolysis reaction of starting halide materials such as $SiCl_4$, $GeCl_4$, $POCl_3$ carried from the supplying system. Such flame hydrolysis reactions are expressed in the following manner:

$$SiCl_4 + 2H_2O \longrightarrow SiO_2 + 4HCl \qquad (4.2)$$

$$GeCl_4 + 2H_2O \longrightarrow GeO_2 + 4HCl . \qquad (4.3)$$

It has been found by modeling hydrolysis reactions in an electric furnace that the reactions in eqs. (4.2) and (4.3) begin taking place in the temperature range of 800 to 900°C and become complete at temperatures higher than 1000°C.[14] The reaction temperature range found in the model experiment is lower by 200-300°C than in the oxidation reaction of the same halide materials.[15] In actual flame hydrolysis, the reaction temperature seems to be lower because of the active atmosphere. The reason why the reaction of the oxide formation in hydrolysis takes place at a lower temperature is thus explained: In the hydrolysis reaction, the SiO_2 product is not formed directly from $SiCl_4$ but rather by passing through the $Si(OH)_4$ state in which it is synthesized by the reaction of $SiCl_4$ with H_2O molecules.[14] Therefore, the practical activation energy for the SiO_2 product formation becomes lower. Furthermore, since in the actual flame hydrolysis reaction the existence of radical O, H, OH ions in the oxy-hydrogen flame seems to promote the formation reaction of the oxide products,[16,17] and intermediate products such as $SiHCl_3$ are formed, the formation reaction of the oxide products takes place in lower temperatures and with higher reaction speeds.[14] At the same time, in the oxidation reaction, higher temperature is required for the formation of oxide products because the oxide products are formed directly from halide materials.[15]

Oxide products synthesized in a manner such as in eqs. (4.2) and (4.3) solidify as fine glass particles and then are deposited onto the surface of the silica rod end. Figure 4.10 shows a photograph of the fine glass particles synthesized by the flame hydrolysis reaction in the VAD process. Particles have a size range from 500 Å to 2000 Å and an almost spherical shape. The particle size is changed by synthesis conditions such as flame temperature and spatial concentration of raw materials. In general, the size of these particles becomes smaller with increasing reaction temperature and increasing spatial concentration. Figure 4.11 shows a photograph of the process of synthesis and

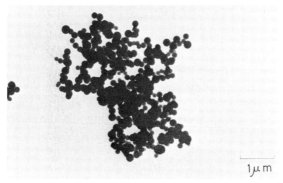

Fig. 4.10. Electron microscope picture of fine glass particles synthesized by flame hydrolysis. [After Sudo[4.57]]

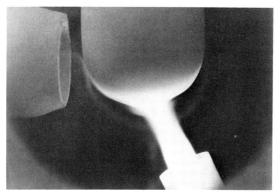

Fig. 4.11. Photograph of the growing surface of the porous preform. [After Sudo[4.57]]

depositon of fine glass particles. It can be seen in this figure that fine glass particles are synthesized in the oxy-hydrogen flame blown from the torch, and then deposited onto the growing surface to form the porous preform, and that H_2O and HCl released by the reaction and residual fine glass particles not attached the growing surface are carried away from the deposition region by the exhausting port.

Figure 4.12 shows a SEM photograph of a porous preform growing surface. It can be seen that fine glass particles make connections among each other to form open networks. Such connections of fine glass particles generate mechanical strength in the porous preform, and resultingly make it possible to prepare the rod-like porous preform in the axial direction. Under ordinary conditions, the density of porous preforms prepared is 0.2–0.5 gr/cm^3 on average,[4] which is 1/10–1/4 times that of silica glass. The mechanical strength of the porous preform is 0.5–1.0 kg/cm^2 on average in the cross sectional area. This strength value provides enough force to pull upward a preform several kilograms in weight.

4.3.2 Preparation and size control of porous preforms

In order to fabricate high-quality fiber preforms, it is of prime importance to prepare rod-like porous preforms having a uniform outer diameter. The stability of the growing surface shape and the uniformity in outer diameter do not depend on the starting silica rod but are almost completely dominated by various deposition conditions such as the flame temperature, vapor supply rate, exhaust gas velocity, preform pulling speed, etc. Therefore, it is of practical importance to stabilize any fluctuations in the system. Fluctuations in the porous preform preparation system are sum-

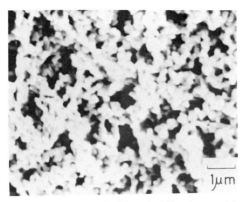

Fig. 4.12. Electron microscope photograph of deposited fine glass particles on porous preform. [After Sudo[4.57]]

marized schematically in Fig. 4.13. Even if both the torch and exhaust port are kept at the fixed locations, fluctuations in the preparation conditions shown in Fig. 4.13 cause undesired variations in the preform outer diameter and in the refractive-index profile. Thus, in order to avoid such fluctuations, it is necessary to use stabilization techniques. This is done in the preparation conditions, starting materials supply rate, and in insuring that the exhaust rate of the reacted gas and the flame temperature are kept at constant values. At the same time, the pulling speed is controlled to keep the growing end of the porous preform at a fixed position by a feedback mechanism which uses a positional signal from a video camera.

Figure 4.14 shows the surface temperature fluctuation measured with a pyrometer at a position near the maximum temperature point of the growing surface for 1 hour. Only a slight fluctuation within ±5°C can be observed. This value is small enough for attaining high-quality fiber preforms. Figure 4.15 shows a photograph of a porous preform prepared using this stabilization technique. Figure 4.16 shows the outer diameter fluctuation of an as-grown transparent preform made by the same technique. The outer diameter is 25 mm and the fluctuation over 200 mm length is ±0.1 mm or ±0.2%. The preform cross section circularity mainly depends on the fluctuation in rotation speed. Figure 4.17 shows the circularity of the as-grown transparent preform. Average diameter is 25 mm and the fluctuation is ±10 μm or ±0.05%.

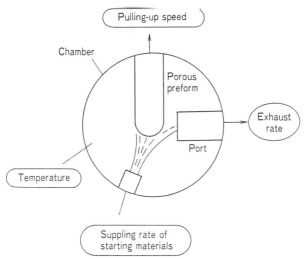

Fig. 4.13. Schematic diagram of fluctuation factors. [After Sudo[4.57]]

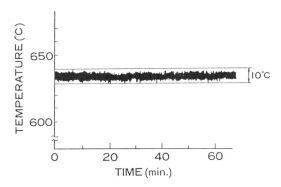

Fig. 4.14. Surface temperature fluctuation measured at a position near the maximum temperature point of the growing surface for 1 hour. [After Sudo[4.57]]

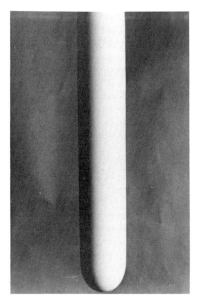

Fig. 4.15. Photograph of a porous preform made by a stabilization techniques. [After Sudo[4.57]]

4.3.3 Simultaneous cladding formation

In order to attain the loss reduction of the VAD optical fiber, it is of importance to simultaneously form a cladding layer having a lower-refractive-index on the periphery of the porous preform during the preparation of the

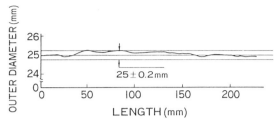

Fig. 4.16. Outer diameter fluctuation of an as-grown transparent preform. [After Izawa et al.[4,5]]

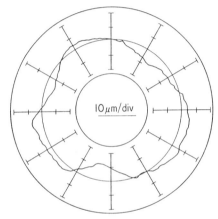

Fig. 4.17. Circularity of an as-grown transparent preform. [After Izawa et al.[4,5]]

core porous preform. It is equally important to prepare the rod-like porous preform without fluctuation in the outer diameter. Figure 4.18 shows a photogrpah of the simultaneous cladding formation; a SiO_2–B_2O_3 layer of fine glass particles is deposited as a cladding on the periphery of a SiO_2–GeO_2 glass porous preform already prepared using another torch. Usually, the cladding glass layer with a thickness half that of the core rod diameter is readily formed with this technique. Therefore, this cladding technique has an extreamly positive effect on the loss reduction in graded-index fibers, step-index fibers and in the high NA fibers. However, a thicker cladding layer is needed for achieving loss reduction in single-mode fibers. This requires that the cladding-to-core diameter ratio is approximately 3 or more in order to fabricate low-loss single-mode fibers.[18]

The above-mentioned porous preform preparation technique can be used to fabricate porous preforms as small as 30 mm in diameter. It is, however,

Fig. 4.18. Photograph of simultaneous cladding formation: An SiO_2-B_2O_3 glass particle layer as cladding is deposited on the periphery of a rod-like SiO_2-GeO_2-P_2O_5 porous glass preform already prepared by the lower torch. [After Sudo[4.57]]

difficult to reduce the diameter of the porous preform to less than 30 mm. If a porous preform having a diameter of 30 mm is used as the porous glass body for the core, the probable cladding-to-core diameter ratio will be approximately 2 at maximum. The reason for this is that if the cladding thickness were increased in this way so as to obtain a ratio of 3 or more, the diameter of the porous preform after cladding would exceed 100 mm at least. The likely result would be that a stress developed therein would possibly crack the porous preform, or the excessively large diameter would render it hard to completely consolidate the porous preform.

Consequently, in order to fabricate a single-mode fiber preform by the VAD method, a porous preform having a small diameter must first be prepared for the core. Next, a thick cladding layer is formed on this slender rod-like porous preform using a subsidiary torch. In addition to the simultaneous cladding technique, a technique for preparing a small-diameter porous preform is absolutely necessary for achieving low-loss VAD single-mode fibers.

A torch suitable for fabricating the small diameter core glass for single-mode fibers has been developed, and its structure is shown in Fig. 4.5(c). The key feature of this torch is that the nozzle for blowing the raw materials deviates from the center of the torch by a distance 1. Figures 4.19(a) and 4.19(b) show schematically the situation of the porous preform preparation for single-mode fiber.[6,9] The core torch is located at an angle with respect to the axial direction of the preform, and the exhaust port is also located near the periphery of the core porous body. The fine glass particle stream from the core torch is blown only in the part position of the frame, and glass particles are

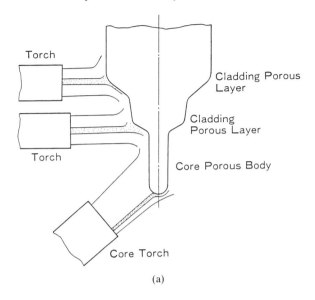

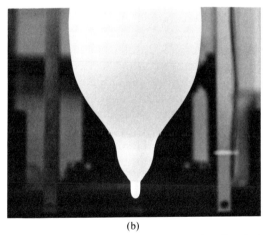

Fig. 4.19. (a) Schematic of the porous preform preparation for single-mode fiber. (b) Photograph of porous preform for single-mode fiber. [After Sudo[4.57]]

deposited only onto the top of the core porous body growing surface. Next, a thick porous glass cladding is formed on the periphery of this core porous body. With this method, a porous preform having a 15 mm diameter core has been made. Using this porous preform, a cladded porous preform with a 75 mm outer diameter can be fabricate even when the cladding-to-core diameter ratio is to be more than 5.

4.3.4 High-speed production of porous preforms

A high-speed production technique for porous preforms has been developed in relation with the cost-reduction of optical fibers.[19,20] Such a technique applies the particle-diffusion effect in the flame stream to the practical fabrication process, based on the investigation of the deposition mechanism of fine glass particles.

Results of deposition experiments suggest that the deposition rate and its efficiency depend on the Reynold's number of the particle stream and have a maximum value of about 30 for high deposition efficiency. It is also found from Schlieren photography that this value of Reynold's number corresponds exactly to the starting value where the eddy of the flame is observed, as in Fig. 4.20. These investigations confirm that particle stream in the flame has an effect on increasing the deposition rate. The particle deposition rate achieved has been reported to be about 4.5 g/min, and large VAD preforms have been successfully produced under these fabrication conditions in which the Reynold's number is set at about 30. Figure 4.21 shows photographs of preforms: (a) A large preform made with this technique, (b) a preform prepared by a conventional method.

4.4 Dehydration and Consolidation

The porous glass preform is consolidated into a bubble-free transparent glass preform by zone sintering using an electric furnace. Dehydration processes are usually included in the consolidation process, where a large amount of OH ions and H_2O molecules contained in the porous preforms are removed. In these sections, the consolidation process is first described, with emphasis on consolidation process analysis, based on the elemental model of closed pores. The important roles of both the environmental gas atmosphere and the rate at which the temperature is increased in the furnace are also presented.

The achievement of a uniform and high degree of transparency in the preform, which is important in relation to transmission loss characteristics for a drawn fiber, is one of the main problems to be solved in the development of the VAD. The consolidation method by zone sintering using the electric furnace has been developed to continuously consolidate the porous preform grown in an axial direction. In the early use of the VAD method, nitrogen or argon gases were used as an environment gas in the consolidation furnace. However, it was difficult to obtain a bubble-free transparent preform, but, by using a helium atmosphere, bubble-free transparent preforms are easily obtained. Accordingly, it seems that the problem of the consolidation process has been solved experimentally. However, the problem of bubbles remaining in the preform has arisen again, along with the development of dehydration techniques and high speed porous preform production. It is, therefore,

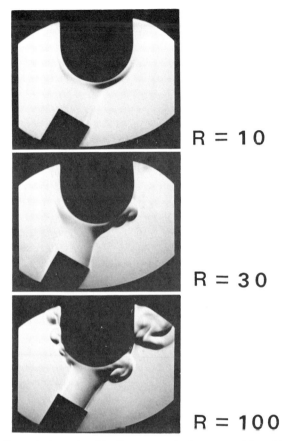

Fig. 4.20. Schlieren photographs of the flame stream in the process of deposition. [After Suda et al.[4.20]]

necessary to completely solve this consolidation process problem, in order to obtain a high-quality VAD preforms.

Sintering of a low density porous preform comprises two processes:[21] densification process of the porous glass, where the transition from open pore state to closed pore state occurs, and the closed pore collapsing process. The former process has been treated by Sherer in relation to the sintering behavior of OVD porous preforms.[22] In his treatment, a cubic array model formed by intersecting cylinders was used as a micro structure of low-density open pore glass preforms; the cubic arrays were densified by viscous flow driven by surface energy reduction. He showed an elegant scheme for porous preform densification. However it is necessary to analize the latter process of sintering

96   Chapter 4

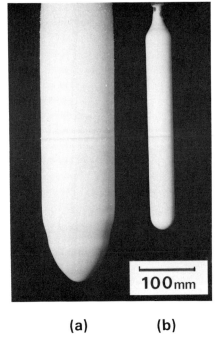

(a)   (b)

Fig. 4.21. Photograph of porous preform: (a) with high-speed production technique, (b) with conventional process. [After Suda et al.[4.20]]

i.e., closed-pore collapsing, which is also important to obtain a highly transparent bubble-free preforms.

This section describes the closed pore behavior during the VAD porous preform zone-sintering process. Present investigation puts emphasis on the role of the environmental gas atmosphere in the furnace. An elementary model, which takes the gas permeation from the closed pore to the surrounding glass wall into consideration, is proposed for the final stage in the sintering process. The experimental results on this model are also detailed.

4.4.1 Experimental results[23,24]

Figure 4.22 shows SEM photographs of a preform at each position from porous state to transparent state, whose positions are illustrated in Fig. 4.23. In this case, the porous preform was sintered in a helium gas atmosphere. Figure 4.22(a) shows the initial porous state: Fine glass particles (0.05–0.2 μm in diameter) form an open network. Figure 4.22(b) shows the first sintering state: Particles begin to melt into each other to form open pores. Gases existing in the open pores can move freely through pores. Figure 4.22(c) presents the second sintering state, where the pores are isolated from each other. At this stage, the preform becomes transparent and the gas is confined in each pore. Porous preform densification is almost complete at this stage. Figure 4.22(d) shows a closed-spherical pore state; the closed pore shapes in the sintered glass are changed because of the low viscosity and surface tension. The final state in the sintering process is shown in Fig. 4.22(e): Pores are collapsed and the preform becomes bubble-free and transparent.

Figure 4.24 shows SEM photographs of a preform sintered in an argon gas atmosphere at each point illustrated in Fig. 4.25. Figure 4.24(a) shows the open pore state which corresponds to Fig. 4.22(b). Figure 4.24(b) indicates the closed pore state. Behavior of the closed pores differs from that in the preform sintered in a helium gas atmosphere: Pores are larger in diameter more than 1 μm, but do tend to expand in diameter. Figure 4.24(c) shows the closed pores in the transparent glass and the expanded pores are shown in Fig. 4.24(d).

Sintering porous preforms in different environmental gas atmospheres leads to different optical quality of the preforms. Table 4.1 shows the glass quality of the preform sintered under the five different atmospheric gas conditions. He to Ar gas ratios are (a) 100:0, (b)80:20, (c) 60:40, (d) 40:60, (e) 20:80, and (f) 0:100 (in mole percent). It can be seen in Table 4.1 that more than 60 mol% He gas content is necessary to sinter a porous preform into a bubble-free transparent preform.

4.4.2 A model for final sintering stage[24]

SEM observations indicate that initial sintering states, from the as-deposited porous state to the closed pore state show a similar trend when sintering is accomplished in He gas atmosphere and in Ar gas atmosphere. The initial sintering state corresponds to the shrinkage process for an as-deposited porous preform, which has been treated theoretically by Sherer.[22] Final sintering states in the two cases (He and Ar atmosphere), however, show quite different behavior: The preform sintered in an Ar gas atmosphere contains closed pores, which are hardly collapsed, as is shown in Fig. 4.24 (c and d). On the other hand, the preform sintered in He gas atmosphere contains no closed pores, since they are completely collapsed during the process. In the following sections, some considerations on the final sintering stage are presented, based on a proposed model. In the model, the different gas permeabilities of the environmental gases in silica glass play an important role in the final sintering stage.

Behavior of the closed pores in the final stage of the sintering process seems to depend on the balance between gas permeation rate into the surrounding glass and pore expansion rate during temperature increase. A simple model illustrated in Fig. 4.26 is proposed; a closed spherical pore filled with inert gas (He and/or Ar) is surrounded by the spherical glass. The glass wall thickness is L and the pore volume is V. On the basis of the pore model,

(a)

(b)

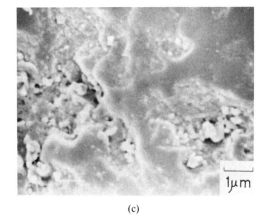

(c)

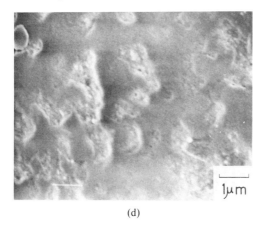

(d)

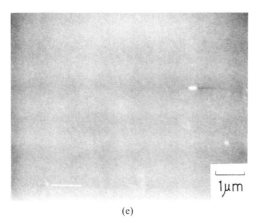

(e)

Fig. 4.22. Electron microscope photograph of the preform at each position from porous state to transparent state sintered in He gas atmosphere. (a) Initial pore state, (b) Open pore state, (c) Closed pore state, (d) Closed-spherical pore state, (e) Bubble-free transparent glass. [After Sudo[4.57]]

pore expansion or shrinkage will be treated as follows.

The relation between pore volume and surrounding temperature can be written by Boyle-Shalle's law

$$PV = nRT \qquad (4.4)$$

where P is the pressure inside the pore, n is mole number for the gas inside the pore and R is gas constant.

Small pore volume increment, ΔV, by a slight temperature increase, ΔT, can be obtained by differentiating eq. (4.4)

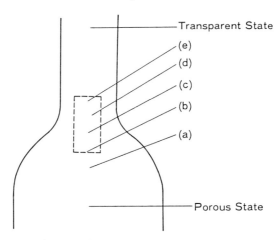

Fig. 4.23. Illustration of each position in the preform corresponding to Fig. 4.22. [After Sudo[4.57]]

$$\Delta PV + P\Delta V = \Delta nRT + nR\Delta T. \tag{4.5}$$

The sign of $\Delta V/\Delta T$, as determined from eq. (4.5), dominates the pore behavior with increasing temperature. The pressure inside the pore, P, can be related to the pressure in the surrounding glass, P_0, by eq. (4.6)

$$P - P_0 = 2\gamma(3V/4\pi)^{-1/3} \tag{4.6}$$

where γ is glass surface tension. The small change in Δn in eq. (4.5) represents mole number variation by the gas diffusion into the glass wall. Δn can be roughly expressed by

$$\Delta n = -JA\Delta t/N_0 \tag{4.7}$$

where J is the gas diffusion density, A is the pore inner surface area, Δt is the time required for increasing the temperature from T to $T+\Delta T$ and N_0 is Avogadro's number. The diffusion density J is approximately given by[25]

$$J = DS(P - P_0)/L \tag{4.8}$$

where D is the diffusion coefficient of the gas in the glass, S is the gas solubility in glass and L is the glass wall thickness. Substituting eqs. (4.6), (4.7) and (4.8) into eq. (4.5) gives

Vapor-Phase Axial Deposition Process

$$\frac{\Delta V}{\Delta T} = \frac{R(n - \beta T)}{P_0 + \frac{4}{3}\left(\frac{4\pi}{3}\right)^{1/3} \gamma V^{-1/3}} \quad (4.9)$$

where

$$n = \frac{V}{RT}\left[P_0 + 2\gamma\left(\frac{4\pi}{3}\right)^{1/3} V^{-1/3}\right] \quad (4.10)$$

$$\beta = \frac{8\pi}{N_0}\left(\frac{3}{4\pi}\right)^{1/3} \frac{\gamma D S V^{1/3}}{CL} \quad (4.11)$$

$$C = \Delta T/\Delta t . \quad (4.12)$$

It is obvious, from eq. (4.9), that the sign of $\Delta V/\Delta T$ changes if $n > \beta T$ or $n < \beta T$. When $n = \beta T$, the pore diameter is given as the critical value,

$$d_c = \left(\frac{3}{4\pi}V_c\right)^{1/3} = -2\left(\frac{36}{\pi^2}\right)^{1/3} \frac{\gamma}{P_0}$$

$$+ \sqrt{4\left(\frac{36}{\pi^2}\right)^{2/3}\left(\frac{\gamma}{P_0}\right)^2 + 48\left(\frac{3}{4\pi}\right)^{1/3} \frac{\gamma R T^2}{P_0 N_0} \frac{DS}{CL}} . \quad (4.13)$$

A pore with larger diameter than d_c expands with the increasing temperature, while a pore with d less than d_c tends to shrink.

It must be noted here that the above mentioned analysis is rather qualitative, because the pore expansion or shrinkage dynamic motion is neglected. In order to analyze the pore behavior dynamic motion, a more complicated consideration is neccessary. It can be said that the glass viscosity does not play an important role on the determination of the pore critical diameter d_c. This may be explained by the fact that the pore with the critical diameter will keep its stationary condition, even if the temperature is increased.

4.4.3 Consolidation condition
A) *Gas permeability effect on pore behavior*
Substituting the following numerical values into eq. (4.13)

$$P_0 = 1 \text{ (atom)}$$

$$\gamma = 3.0 \times 10^{-4} \text{ (atom·cm)}[26)]$$

$$R = 82 \text{ (cm}^2\text{·atom·K}^{-1}\text{·mol}^{-1})$$

$$T = 1600 \text{ (K)} .$$

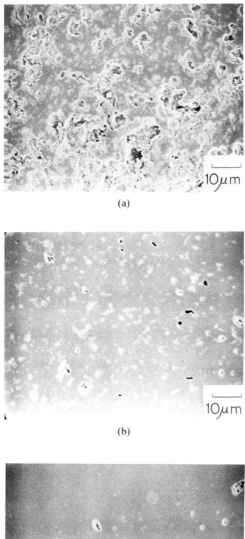

(a)

(b)

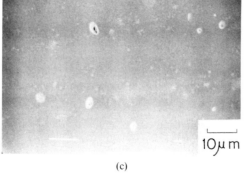

(c)

Vapor-Phase Axial Deposition Process

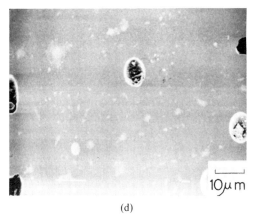

(d)

Fig. 4.24. Electron microscope photograph of the preform at each position from porous state to transparent state sintered in Ar gas atmosphere. (a) Initial pore state, (b) Open pore state, (c) Closed pore state, (d) Closed-spherical pore state. [After Sudo[4.57]]

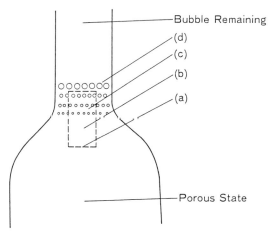

Fig. 4.25. Illustration of each position in the preform corresponding to Fig. 4.24. [After Sudo[4.57]]

We obtain the critical pore diameter

$$d_c = -0.545 \times 10^{-3} + (0.297 \times 10^{-6} + 3.09 \times 10^2 K/CL)^{1/2} \qquad (4.14)$$

where $K=DS$ and is the gas permeability in glass. The critical diameter d_c

Table 4.1. The glass quality of the preform sintered under the five different atmospheric gas conditions.

He:Ar RATIO (mol %)	GLASS QUALITY
100:0	Bubble free
80:20	Bubble free
60:40	Bubble free
40:60	Slightly bubble remaining
20:80	Bubble remaining
0:100	Bubble remaining

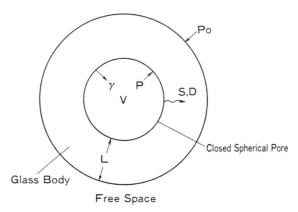

Fig. 4.26. A model of the closed-spherical pore in the final sintering stage. [After Sudo[4.57]]

increases with increasing gas permeability, K, and decreases with increasing temperature speed, C, and wall thickness, L.

The temperature increasing speed, C, in the present experimental condition was about 1 K/sec, estimated from the temperature gradient (140°C/cm) in the furnace and the preform pulling speed (4 mm/min). The wall thickness, L, was obtained from the SEM photograph at the transition as about 0.1 cm, though the situation at the transition zone shown in Fig. 4.22 is more complicated than the simple case shown in Fig. 4.26. Permeabilities for He and Ar in silica glass, using data by Perkins,[25] were found to be

$$K_{He} = 8.32 \times 10^{-7} \text{ (cc(STP)} \cdot \text{cm}^{-1} \cdot \text{sec}^{-1} \cdot \text{atm}^{-1} \cdot \text{K}^{-1})$$

$$K_{Ar} = 2.27 \times 10^{-11} \text{ (cc(STP)} \cdot \text{cm}^{-1} \cdot \text{sec}^{-1} \cdot \text{atm}^{-1} \cdot \text{K}^{-1}) .$$

Substituting these data into eq. (4.8), the critical diameters for He and Ar result

$$d_{c,He} = 500 \ \mu m$$

$$d_{c,Ar} = 0.6 \ \mu m \ .$$

This means that the Ar gas filled pores, with larger than 0.6 μm pore diameter, do not shrink with the increasing temperature. The present observation shown in Fig. 4.24 indicates a fairly good agreement with the estimated critical diameter scheme. In the case of a He gas atmosphere, on the other hand, the closed pores with a smaller diameter than 500 μm are rarely formed in the usual densification process during the initial sintering process. Figur 4.27 shows the relation between the gas permeability and the critical diameter increase with the increasing gas permeability. The bubble-free transparent preform is easily obtained in atmosphere containing a more highly permeable gas.

B) *Temperature increasing speed*

Obviously, from eq. (4.13) or (4.14), the critical diameter, d_c, increases with decrease in temperature increasing speed, C. This result indicates that in the zone sintering case, a transparent preform is more easily obtained when the temperature gradient at the hot zone region along its axis decreases and also the pulling speed decreases. In the uniform sintering case, the transparent preform is mor easily obtained when the temperature increasing speed is slow.

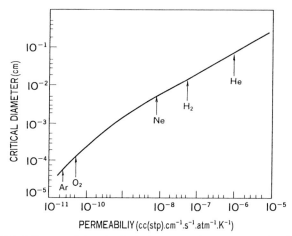

Fig. 4.27. Critical diameter of closed-spherical pore to make transparent glass in various kinds of gases having differnt permeability. [After Sudo[4.57]]

C) *Sintering in mixed gas atmosphere*

Table 4.1 shows some examples of preform transparency, when sintered under different gas conditions. The result indicates that it is necessary to maintain a sintering atmosphere with more than 60% He gas to obtain bubble-free preforms. This result can be explained as follows, based on the above mentioned discussion.

Gas permeability K_m for an He-Ar mixed gas is presented as

$$K_m = K_{He} M_{He} + K_{Ar} M_{Ar} \tag{4.15}$$

where M_{He} and M_{Ar} are the mixed ratio of He and Ar, respectively. The critical pore diameter, where the pore starts to shrink, is given by substituting eq. (4.15) into eq. (4.14), giving

$$d_{cm} = -0.545 \times 10^{-3} + [0.297 \times 10^2 \, K_m/CL]^{1/2} . \tag{4.16}$$

This equation indicates that a closed pore, with diameter, d, less than d_{cm}, starts to shrink until most of the He gas diffuse out from the pore into the surrounding glass. As a result, the closed pore containing residual Ar gas has a diameter given by $d M_{Ar}^{-1/3}$. So, if this pore diameter is smaller than the critical diameter $d_{c,Ar}$, that is, if the initial pore diameter is smaller than $d_{c,Ar} M_{Ar}^{-1/3}$, the closed pore tends to diminish. In case of $M_{Ar} = 0.4$, the value of $d_{c,Ar} M_{Ar}^{-1/3}$ is about 0.8 μm and d_{cm} is about 120 μm. The above discussion shows that a pore with smaller than 0.8 μm diameter will finally diminish. On the other hand, a pore with between 120 μm and 0.8 μm diameter only shrinks until He gas diffuses out and the pore does not diminish. Experimental results, shown in Table 4.1, indicate that actual preforms mostly contain closed pores with smaller diameter than 0.8 μm, though the above discussion is based on a rather oversimplified consideration.

It seems worthwhile to note finally that the discussion in this section is of practical interest in connection with the recent developments in the dehydration technique for the porous preforms: Mixtures containing halide gas are used during the sintering process, so as to eliminate OH ions and H_2O molecules from the preforms, as mentioned in Section 4.5.

It must also be noted here that the critical diameter depends not only on gas permeability K but also on temperature increasing speed, C. It was found that a bubble-free preform can be obtained, even in pure Ar gas atmosphere, with a temperature increasing speed smaller than 0.1°C/sec.

The above results are not only useful for determining the sintering condition for porous preforms and designing the furnace structure for sintering process in the VAD method, but also are applicable to the OVD sintering process.

4.5 Dehydration

In this section, the detailed dehydration techniques, using $SOCl_2$ and Cl_2 gases, technical points for achieving complete dehydration and the dehydration equipment, are described. Progress in the dehydration-consolidation techniques and the resulting reduction in residual OH-ion content are finally presented. The VAD porous preform consists of fine glass particles, of sizes 500–2000 Å in diameter, as shown in Fig. 4.10. Since fine glass particles are synthesized as a result of flame hydrolysis reactions, these particles contain a large amount of OH ions and water molecules in and/or on the particle surface, which then contaminate the transparent preform. Accordingly, in order to realize the loss reduction, especially in the long wavelength region, it is necessary to effectively remove residual OH ions and water molecules.

4.5.1 Dehydration principle

Particle structures containing absorbed water molecules and OH ions are illustrated in Fig. 4.28, which is predicted from the observed data.[27] The process of removing such OH ions and water molecules has long been studied, from the viewpoint of the surface chemistry.[28-30] Its behavior is qualitatively interpreted as follows: First, when fine glass particles containing OH ions and

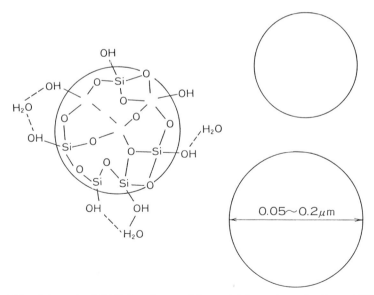

Fig. 4.28. Schematic of VAD fine glass particles containing water molecules and OH ions. [After Sudo[4.57]]

water molecules are heated in a dry atmosphere, the physically absorbed water molecules are readily removed at 150°C and then part of the chemically bonded Si–OH are dehydrated at about 400°C: However, some Si–OH bonds remain on the particle surface, as isolated OH ions, even at an elevated temperature of about 800°C as illustrated in Fig. 4.29.[27,31] The OH ions in the particles diffuse out according to the OH ion concentration gradient and diffused-out OH ions condense to water molecules. Even so, isolated OH ions remain on the surface. So, the porous preform consolidation in dried atmosphere results in some amount of residual OH ions in the transparent preform (OH content is experimentally 5–30 ppm). If we consider that one or two ions stays in a 10 nm^2 area on the particle surface whose particle size is 0.1 μm in diameter on the average,[32] it can be estimated as 30 ppm residual OH ions comparable to the above experimental value. In order to reduce such isolated OH ions, chemical treatment using a chemical reactive reagent is necessary.

Thionyl chloride, $SOCl_2$, was mainly used as a dehydration reagent, because its chemical properties and dehydration effect on silica glass surface are well known and its reactivity as a dehydration reagent is very effective. The chemical reaction of $SOCl_2$ with H_2O and OH ions are expressed by the following:[32,34]

$$SOCl_2 + H_2O \longrightarrow SO_2 + 2HCl \qquad (4.17)$$

$$SOCl_2 + Si\text{–}OH \longrightarrow SO_2 + HCl + Si\text{–}Cl . \qquad (4.18)$$

This chemical treatment is essentially the halogenating process, where an isolated OH ion is replaced by Cl-, Br- or F-ions. This halogenating process resultingly generates Si–Cl bonds: However, absorption loss due to such Si–Cl bond has no serious influence on the fiber loss in the wavelength region of present interest, because the fundamental absorption peak due to Si–Cl bond is around 25 μm wavelength. Thus, the chemical dehydration utilizing the halogenating process has an important effect on the reduction in residual OH-ion content.

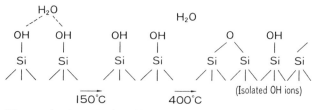

Fig. 4.29. Diagram showing behavior of H_2O molecules and OH ions on the particle surface. [After Sudo[4.57]]

4.5.2 Optimum dehydration temperature[31,35]

The chemical dehydration reaction is sensitive to the furnace temperature. Figure 4.30 shows the relation between the dehydration temperature and the residual-ion content in consolidated preforms. In the experiment, only the dehydration temperature was changed using $SOCl_2$ as a dehydration reagent and the uniform-heating furnace. The dehydration treatment was carried out for 2 hours.

It can be seen in Fig. 4.30 that the residual OH-ion content steeply decreases in the vicinity of 700°C and about 1200°C. About 30 ppm OH ions remain in the preform treated at temperatures lower than 700°C and about 0.5 ppm OH ions remain in preforms treated at the temperature range between 700°C and 1200°C. Lastly, OH ions could be reduced to less than 0.1 ppm level by treating the porous preform at temperatures higher than 1200°C.

This experimental result can be explained as follows. First, a critical point appears at 700°C in Fig. 4.30, where the residual OH content decreases sharply, and is limited by the chemical reaction between OH ions and $SOCl_2$ on the particle surface,[36] rather than OH diffusion. This is so because the OH diffusion length, $(4DT)^{1/2}$, in silica glass, calculated using the diffusion constant for OH ions in silica glass at 600°C ($D=10^{-11}$ cm^2/sec) and dehydration time, is about 505 μm, which is larger than particle size. It indicates that most of the OH ions in the particle volume easily diffuse out to the particle surface. The porous preform starts to shrink at about 1200°C.

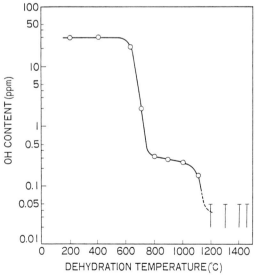

Fig. 4.30. Relation between dehydration temperature and residual OH ion content. [After Sudo[4.57]]

However, until consolidation finishes, preforms contain many pores, as can be seen in Fig. 4.22, water vapor easily migrates into the open pores and recombines with or adheres to the particle surface by the recombination process from Si–Cl to Si–OH. It means that dehydrated particle surfaces are recombined with OH ions or H_2O molecules during the consolidation period, when water vapor is contained in the environment. It also means that it is necessary to maintain a water vapor free atmosphere, even during the consolidation period. Gases like helium or oxygen used during consolidation usually contain more than 1 ppm water. To prevent further contamination during the consolidated process, it is necessary to maintain a flow of a dehydration reagent, like chlorine or thionyl chloride vapor, into the furnace throughout the dehydration period and consolidation process.

4.5.3 Water vapor pressure[31,35]

Figure 4.30 shows the dehydration effect of $SOCl_2$ on a porous preform. The residual OH contents in a preform, consolidated in an atmosphere containing water vapor, are plotted versus the square root of water vapor pressure. The open circles indicate the OH concentration data in the preform consolidated without $SOCl_2$ and solid circles with $SOCl_2$. The vapor pressure of $SOCl_2$ was set at 10 mmHg. The broken line in the figure represents the estimated vapor pressure using water solubility in silica glass.

In the high water vapor pressure region, measured and estimated values coincide well with each other. Residual OH content in the preform consolidated without $SOCl_2$ increases almost linearly, according to the increase in water vapor pressure. On the other hand, in the lower vapor pressure region, some difference appears between experimental and estimated values. Even at zero water pressure 300 ppm OH ions remain in the consolidated preform. When 10 mmHg $SOCl_2$ partial pressure was carried into the furnace saturated with oxygen gas, residual OH ion content shows a steep increase in the 10 mmHg water vapor pressure point vicinity.

This experimental result, shown in Fig. 4.30, indicates that water vapor pressure in the atmosphere should be reduced to as low a level as possible, in order to successfully reduce the residual OH-ion content.

4.5.4 Dehydration reagent pressure[35]

Figure 4.31 shows the relation between the partial pressure for the dehydration reagent (Cl_2 gas) in the atmosphere and residual OH-ion content. It can be seen in the figure that the residual OH-ion content reduces with increase in the Cl_2 gas partial pressure. It is, however, necessary to choose an adequate partial pressure, taking into account the critical diameter given by eq. (4.16). The result shown in Fig. 4.32 and the restriction in the consolidation suggest that the adequate partial pressure ranges from 10 mmHg to 100 mmHg.

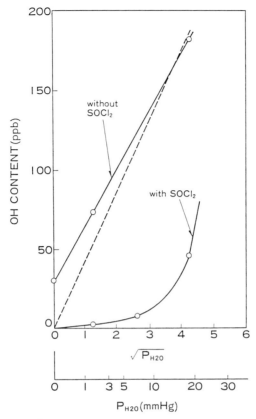

Fig. 4.31. Residual OH-ion content in the preforms treated in humid atmosphere with and without SOCl$_2$. [After Sudo[4.57]]

4.5.5 Dehydration time

Figure 4.33 shows the relation between the dehydration time and the residual OH-ion content obtained under condition of 770°C dehydration temperature and 10 mmHg SOCl$_2$ partial pressure. It can be seen that the residual OH content is reduced to a low level, less than 0.4 ppm, by a dehydration time longer than 2 hours. This result means that the dehydration time is not determined by the OH-ion diffusion time from the particle volume, but by the speed of the SOCl$_2$ reaction with OH ions on the particle surface. Of course, higher partial pressure for SOCl$_2$ gas enable shorter dehydration times. However, 2 hours dehydration time is necessary to reduce residual OH content to the 0.01 ppm level or less.

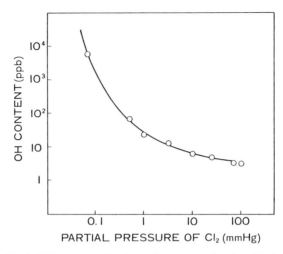

Fig. 4.32. Residual OH-ion content in the preforms treated under chlorine gas containing atmosphere. [After Sudo[4.57]]

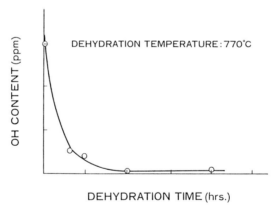

Fig. 4.33. Treating time dependence of residual OH-ion content in the chlorine-treated preform. [After Sudo[4.57]]

4.5.6 Kinds of dehydration reagent

As mentioned above, chlorinating reagents, such as $SOCl_2$ or Cl_2, have been mainly used as the dehydration reagent. Other halogenating reagents, such as $SOBr_2$ (brominating reagent) and $C_2Cl_2F_2$ (fluorinating reagent), were also investigated.

Figure 4.34 shows a comparison between the infrared transmission spectra for the transparent preform dehydrated with $SOCl_2$ and that for the

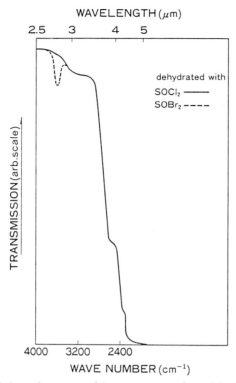

Fig. 4.34. Infrared absorption spectra of the transparent preforms dehydrated with $SOCl_2$ and $SOBr_2$. [After Sudo[4.57]]

transparent preform dehydrated with $SOBr_2$. It is found that residual OH content in the transparent preforms dehydrated with $SOBr_2$ is one order of magnitude larger than that dehydrated with $SOCl_2$. This result means that the Cl-ion reactivity with Si–OH is stronger than the Br-ion reactivity with Si–OH.[27] This is understood in terms of the difference between Cl atom and Br atom in electronegativity defined by L. Pauling.

When a fluorinating reagent, whose dehydration reactivity seems to be the strongest of various halogenating reagents, is used, the silica glass particles are fluorinated in themselves and vaporized. Therefore, a chemical dehydration treatment using the fluorinating reagent is basically difficult specially in the case of wet porous preforms. However, fluorination is used for making glass with a lower refractive-index than that of pure silica and for the termination of dangling bonds to prevent the effects of hydrogen permeation.[37] The details are mentioned in Section 4.6.5.

4.5.7 Cladding thickness

OH ions easily diffuse into the preform during the elongation period for adjusting the core/cladding diameter ratio to the fiber specification by jacketting with a silica tube. The thick cladding layer deposition is necessary to achieve a completely OH free fiber, in addition to the dehydration process. Figure 4.35 shows the relation between the SiO_2 cladding layer thickness deposited simultaneously and the absorption loss due to OH ions in the drawn fiber at 1.39 μm wavelength. The preform is dehydrated and consolidated under conditions in which the dehydration temperature is higher than 1200°C, the Cl_2 gas partial pressure is 100 mmHg and the dehydration time is 2 hours. It can be seen that the residual OH content is successfully reduced to a level less than 10 ppb, considering the corresponding relation of the 0.6 dB/km OH-absorption loss at 1.39 μm to the 10 ppb OH-ion content. Further, completely OH free fibers are achieved by using an all VAD synthesized fiber:[38] Its residual OH-ion content is less than 1 ppb.

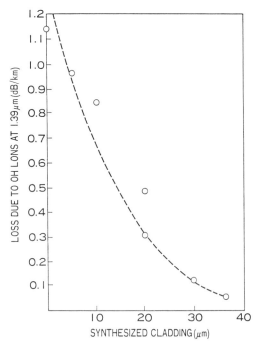

Fig. 4.35. Relation between synthesized cladding glass thickness and absorption loss due to residual OH ions in outer cladding glass. [After Sudo[4.57]]

4.6 Profile Control

4.6.1 Introduction

The key feature of the VAD method is that a porous preform is prepared in the axial direction. This results in refractive-index profile control in the spatial domain, which is not the case with the MCVD and OVD methods.

In the early stages of development of the VAD process, the profile formation concept was determined on the basis that raw material vapors for different compositions, being blown out from different torch nozzles, mixed spatially with each other by diffusion and overlapped to form a spatial dopant distribution on the growing surface of the porous preform.[1-5,39] This idea was thought to be dominant in profile formation. However, the fact that graded-index profiles could be made through the use of even a single torch with a single raw material vapor nozzle could not be explained by this mixing effect idea only.[40]

Precise experiment of SiO_2–GeO_2 particle deposition properties in flame hydrolysis reactions has clarified that the GeO_2 concentration in the deposited particle depends strongly on the substrate temperature during deposition. There is, however, difficulty in fully clarifying the relation of these deposition properties to practical fabrication conditions for graded-index fibers. This is because of the complicated behavior shown by flame hydrolysis reaction and synthesized glass particle deposition.

This section is meant to clarify in detail the mechanism underlying profile formation in the VAD process, and to present precise control techniques for graded-index profiles. The emphasis of the description firstly clarifies details of the flame hydrolysis reaction. Secondly the results of the clarification are applied to practical control techniques over the refractive-index profile.

Vapor-phase reactions in the flame and on the porous preform growing surface has a direct influence on the index profile formation. This is because the dopant concentration distribution is spatially formed on the growing surface as a result of flame hydrolysis of the halide raw materials. In order to establish a precise control technique for the refractive-index profile, it is necessary to investigate details of the vapor-phase reaction.

The purpose of the present section is to elucidate the vapor-phase reaction in this flame. In what follows, three experiments will be reported, in which the vapor-phase reactions in the flame and on the porous preform growing surface were investigated. The profile formation mechanism will also be clarified with these experiments.

4.6.2 Dopant concentration[41-43]

With VAD, fine glass particles are deposited on a porous preform surface, which is directly bleached in the oxy-hydrogen flame containing the raw halide materials. It will be shown through the following experiment that dopant concentration and phase of fine glass particles produced in the flame of the synthesizing torch depend strongly on the porous surface temperature.

Investigation of infrared absorption spectra and X-ray diffraction patterns for fine glass particles prepared at different substrate temperatures will show the dopant concentration and the phase. Figure 4.36 shows the experimental arrangement used for the deposition of oxide particles.[41] A vapor-phase binary mixture of $SiCl_4$–$GeCl_4$, $SiCl_4$–BBr_3 or $SiCl_4$–$TiCl_4$ was fed into an oxy-hydrogen torch and the oxide particles synthesized in the flame were deposited onto a silica tube. The surface temperature of the substrate tube was maintained at the desired value between 200°C and 800°C by introducing a cooling gas into the tube. The temperature was monitored using a two-dimensional optical pyrometer.

Infrared spectra for SiO_2–GeO_2 products have strong absorption peaks at 870 cm^{-1} and 660 cm^{-1}, which could be attributed to the vibration of Ge–O–Ge and Si–O–Ge chains, respectively.[44] It suggests that there is an interconnected structure between SiO_2 and GeO_2 other than the simple oxides of Si and Ge. Absorption peaks at 1380 cm^{-1}, 920 cm^{-1} and 950 cm^{-1} are attributed to the vibration of B–O–B, Si–O–B and Si–O–Ti chains, respectively.[45,46] The relative dopant concentration of fine glass particles prepared at different substrate temperature can be determined from these absorption peaks. The phase, whether crystalline or amorphous, of the fine particles is determined from X-ray diffraction patterns.[47] Sharp diffraction peaks which are attributed to the crystalline state of GeO_2 and B_2O_3 are observed for the samples prepared at lower temperatures.

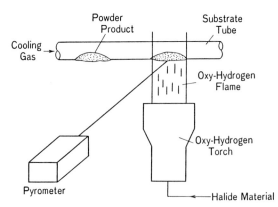

Fig. 4.36. Experimental arrangement for measuring the deposition rate dependence on the surface temperature. [After Sudo[4.57]]

These experimental results confirmed that the SiO_2–B_2O_3 and SiO_2–GeO_2 systems are similar in deposition properties. The concentration and the crystalline structure show a remarkable dependence on the substrate temperature. In the SiO_2–TiO_2 system, in contrast, no significant dependence of TiO_2 concentration on the substrate temperature was found. Figures 4.37, 4.38 and 4.39 schematically illustrate the substrate temperature dependence of GeO_2, B_2O_3 and TiO_2 concentrations, respectively, as evaluated from the IR and X-ray data. The GeO_2 component in the SiO_2–GeO_2 system was deposited with a crystalline structure at substrate temperatures below 400°C, but assumed a noncrystalline form dissolved in the SiO_2 glass network about 500°C. The crystalline GeO_2 concentration decreases with increasing substrate temperature. Dissolved GeO_2 concentration increases with increase of the substrate temperature from 500°C to 800°C.

The low-temperature phase of the B_2O_3 component, formed in the low substrate temperature region below 500°C (Fig. 4.38), is a crystalline state combined with some water as an impurity-hydrate form. The crystalline B_2O_3 concentration decreases with increasing substrate temperature. The dissolved B_2O_3 concentration increases with an increase in substrate temperature from 200°C to 800°C. The data in Figs. 4.37 and 4.38 indicate that the formation of SiO_2–B_2O_3 and SiO_2–GeO_2 particles is not complete in the flame. The

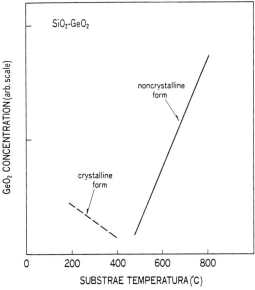

Fig. 4.37. Substrate temperature dependence of GeO_2 concentration in the SiO_2-GeO_2 system. [After Sudo[4.57]]

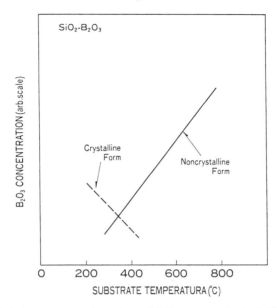

Fig. 4.38. Substrate temperature dependence of B_2O_3 concentration in the SiO_2-B_2O_3 system. [After Sudo[4.57]]

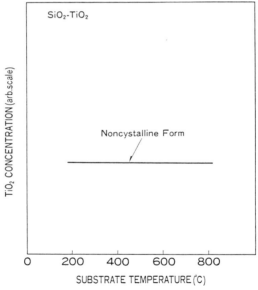

Fig. 4.39. Substrate temperature dependence of TiO_2 concentration in the SiO_2-TiO_2 system. [After Sudo[4.57]]

dependence of the GeO_2 and B_2O_3 concentrations and their crystalline structures on the substrate temperature suggests that the solidification of the GeO_2 and B_2O_3 components occurs near the surface of the substrate, depending on the substrate temperature. GeO_2 and B_2O_3 seem to remain in the vapor state in the oxy-hydrogen flame.

The deposition properties of the SiO_2–TiO_2 system are in sharp contrast to those of the SiO_2–GeO_2 and SiO_2–B_2O_3 systems. The TiO_2 component is deposited with a noncrystalline structure dissolved in SiO_2 glass, irrespective of the substrate temperature, and the TiO_2 concentration does not depend significantly on the substrate temperature.

It is worth noting that the saturated vapor pressures of GeO_2 and B_2O_3 at high temperatures of about 1200–1500°C are several orders of magnitude larger than the values for SiO_2 and TiO_2.[48] This seems to have an important effect on the deposition properties of high silica glass particles in the hydrolysis reaction. In the SiO_2–TiO_2 system, TiO_2 vapor produced in the flame will immediately solidify to make an interconnected structure with co-existing SiO_2 vapor because of their low saturated vapor pressures in the flame. The formation of the SiO_2–TiO_2 particles is, therefore, completed in the flame. This leads to the independence of the TiO_2 concentration on the substrate temperature.

In the SiO_2–GeO_2 system, GeO_2 produced in the flame remains in the vapor state because of its large saturated vapor pressure, and only SiO_2 forms solid particles in the flame. The vapor-phase GeO_2 component is carried with the SiO_2 particles to the surface of the substrate and cooled on the surface. When the substrate temperature is sufficiently low, the vapor-phase GeO_2 solidifies in crystalline form on the substrate. When the substrate temperature is higher, the vapor-phase GeO_2 component is not cooled sufficiently on the surface to be deposited in a crystalline state. Figure 4.37 indicates that at higher temperatures, above 500°C, the GeO_2 component could remain on the substrate, thus forming an interconnected structure with SiO_2. Figure 4.39 schematically illustrates possible particle structures for the SiO_2–GeO_2 system.[43] These structural models are based on the assumption that the solidification of the GeO_2 component is proceeded by formation of SiO_2 particles in the flame, as discussed above. Figure 4.40(a) represents a structural model for the products prepared at lower substrate temperatures below 400°C. The crystalline GeO_2 portion is not covered with a noncrystalline SiO_2 layer, but exists outside the SiO_2 particles. Figure 4.40(b) is a structural model for the products prepared at higher substrate temperatures above 500°C; the central part of the particles is a noncrystalline SiO_2 phase covered with noncrystalline SiO_2–GeO_2 phase.

The validity of the structural models presented in Fig. 4.40 are verified with simple experiments.[43] The SiO_2–GeO_2 product, made with a substrate temperature of 200°C, was dispersed in hot water (80°C). After 10 minutes,

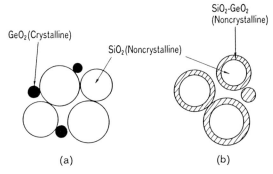

Fig. 4.40. Particle structure models for SiO_2-GeO_2 products. (a): the particles made a temperature lower than 400°C, (b): made higher than 500°C. [After Sudo[4.57]]

the solid particles were separated from the solution by filtration. Figure 4.41 shows X-ray diffraction patterns for the original particles, and particles after filtration. Curve (c) in the figure shows an X-ray diffraction pattern for the solid product obtained by drying the filtrate. Pattern (a) consists of broad

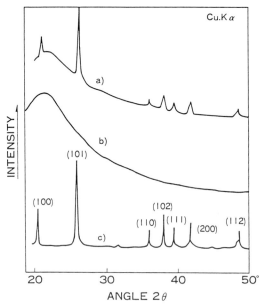

Fig. 4.41. Changes in X-ray diffraction pattern of SiO_2-GeO_2 products made at the surface temperature of 200°C. a) original product; b) residual particles after filteration; c) product obtained by drying filtrate. [After Sudo[4.57]]

peak at $2\theta=22°$ (noncrystalline SiO_2 phase) and several sharp peaks (hexagonal crystalline GeO_2 phase), as was expected. Pattern (b) shows only a broad diffraction peak corresponding to the pure SiO_2 phase. Pattern (c), on the other hand, has only sharp peaks corresponding to pure hexagonal crystalline GeO_2 phase.

It is interesting to note here that hexagonal crystalline GeO_2 is soluble in hot water.[49] Figure 4.41 indicates that the hexagonal crystalline GeO_2 is not covered with the noncrystalline SiO_2 phase (insoluble in water). This supports the validity of structural model of Fig. 4.40(a) for SiO_2 GeO_2 products prepared at substrate temperatures below 400°C.

The SiO_2–GeO_2 product prepared at a substrate temperature of 700°C was dispersed in an etching solution containing 5 wt% $(NH_4)_3PO_4$ and 30 wt% NH_4F. This solution exhibited a slow etching rate of about 20 Å/min. After 15 minutes, the particles were separated from the solution by filtration, washed with water, and dried under a vacuum. Curves (a) and (b) in Fig. 4.42 respectively show IR spectra before and after the etching procedure. It can be seen from the figure that Si–O–Ge absorption, seen in spectrum (a) at 660

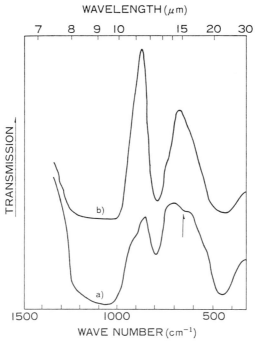

Fig. 4.42. Changes in infrared spectrum of SiO_2-GeO_2 product made at the surface temperature of 700°C. a) before etching the particle surface; b) after etching. [After Sudo[4.57]]

cm^{-1}, disappears in spectrum (b). This supports the structural model in Fig. 4.40(b), where the noncrystalline SiO_2 phase is covered with a noncrystalline SiO_2–GeO_2 phase.

4.6.3 Profile formation mechanism

Various experimental results have been presented in an attempt to elucidate the vapor-phase reaction in the VAD process.[41-43,49,51-53] It is conceivable that these results are closely related to the profile formation mechanism in VAD, particularly as to whether the dopant concentration in the deposited particles depends on the substrate temperature or not. Thus, the vapor-phase reaction process for each glass composition in the flame hydrolysis will be interpreted based on these results. Emphasis will be placed on the presence of substrate temperature dependence of dopant concentration.

A) SiO_2–TiO_2 system

TiO_2 vapor produced in the flame will immediately solidify and make an interconnected structure with co-existing SiO_2 vapor because of low saturated vapor pressures in the flame. The formation of SiO_2–TiO_2 glass particles is, therefore, completed in the flame. This leads to the fact that the TiO_2 concentration and its crystalline structure do not depend on the substrate temperature.

B) SiO_2–GeO_2 system

In this system, only the SiO_2 component is formed as solid particles in the flame, while the GeO_2 product remains in the vapor state. Vapor-phase GeO_2 will be carried to the substrate (actually, the porous preform) surface and be cooled there. When the substrate temperature is sufficiently low, the vapor-phase GeO_2 will solidify in crystalline form on the substrate. On the other hand, when the substrate temperature is higher, vapor-phase GeO_2 is not cooled sufficiently quickly on the substrate for it to be deposited in a crystalline state. As indicated in Section 4.6.2, at higher substrate temperatures, above 500°C, the GeO_2 component can remain in the substrate, forming an interconnected structure with SiO_2.

This GeO_2 component interconnected with SiO_2 is formed by the reaction of GeO_2 vapor with raw $SiCl_4$ gas on the surface of the SiO_2 particles already formed in the flame. Dissolved SiO_2–GeO_2 particles are consequently formed. In this structure, the inner part is a noncrystalline SiO_2 phase covered with dissolved SiO_2–GeO_2 phase. The formation of the GeO_2 component interconnected with SiO_2 also depends largely on both the reaction temperature (i.e., the substrate temperature) and environmental gas composition. Accordingly, the GeO_2 concentration in the SiO_2–GeO_2 glass particles deposited on the substrate surface shows substrate temperature dependence.

Based on these results concerning the vapor-phase reaction in the flame hydrolysis, the formation mechanism of the refractive-index profile, that is the dopant concentration distribution, can be explained as follows. When the porous preform is grown by the deposition of fine glass particle, the parabolic temperature distribution, where the center part is at a high temperature with a gradually lowering of temperature in radial direction, is usually formed on the porous preform growing surface as seen in Fig. 4.43. The substrate-temperature dependence of the GeO_2 concentration plays an important role in the index-profile formation. In the SiO_2–GeO_2 system having the deposition property shown in Fig. 4.37, when the porous preform is grown under surface temperatures ranging from 500°C to 800°C, the GeO_2 concentration gradient of the central part is high with a gradually decreasing concentration to zero in the radial direction. This is automatically formed in the core porous preform according to the surface-temperature gradient. In the SiO_2–TiO_2 system, however, since the TiO_2 concentration has no substrate temperature dependence, the step-like TiO_2 concentration will be formed in the core porous preform, though the same surface temperature gradient as in the SiO_2–GeO_2 system is formed. This step-like TiO_2 concentration has been confirmed experimentally.

The experimental results presented in this section indicate that in the VAD process the temperature gradient of the porous preform growing surface is one of the most important tools for achieving fine control of the refractive-index profile. In the actual VAD process, however, the index profile is not only controlled by the temperature gradient. Other factors, such as the flame temperature and the spatial dopant distribution in the flame, also play important roles in profile formation. Details for other factors concerning profile formation in the VAD process will be presented in the following section.

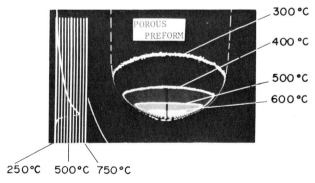

Fig. 4.43. Surface temperature distribution of the porous preform in the process of deposition. [After Sudo[4.57]]

C) Flame temperature effect

It is to be noted that not only the surface temperature, but the flame temperature also has a significant influence on the GeO_2 concentration. Figure 4.44 shows a contour line map for GeO_2 concentration in the SiO_2–GeO_2 system vs. flame and substrate temperature. This result is obtained by using the same experimental apparatus as shown in Fig. 4.36 and changing the flame temperature. The flame temperature was measured by inserting a thermocouple directly into the flame. Therefore the temperatures are the reactive temperature. The numbers in this figure represent the GeO_2 concentration in molar percent. It is seen in Fig. 4.44 that the GeO_2 concentration depends not only on substrate temperature but also on flame temperature. Under fixed flame temperature conditions of 1200°C and 1300°C, the GeO_2 concentration increases with increasing substrate temperature, while under a high fixed flame temperature condition of 1400°C, the GeO_2 concentration tends to decrease at substrate temperatures higher than 700°C. The GeO_2 concentration decreases in the high flame and substrate temperature region in Fig. 4.44 might be attributed to the fact that the GeO_2 product is difficult to deposit at a high temperature as it remains as a vapor because of its high saturated vapor pressure in the flame.

In the case of the SiO_2–TiO_2 system, neither a flame temperature dependence nor a substrate temperature dependence of TiO_2 concentration was found. This is in sharp contrast to the case of the SiO_2–GeO_2 system; the reason could be attributed to the difference in the saturated vapor pressure values between GeO_2 and TiO_2.

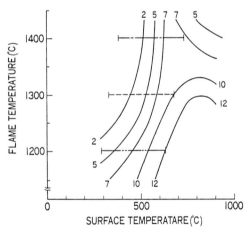

Fig. 4.44. Contour line map for GeO_2 concentration in SiO_2–GeO_2 system shown as a function of flame temperature and substrate temperature. The numerals represent GeO_2 concentration in molar percent. [After Sudo[4.57]]

The result in Fig. 4.44 indicates that care must be taken, both in regard to the flame and substrate temperatures during deposition for obtaining a desired profile in case of the SiO_2–GeO_2 system.

D) Gas mixing effect

In order to achieve precise control over the refractive-index profile in the VAD process, it is necessary to clarify not only temperature effects but also gas mixing effect. Figure 4.45 is a drawing of the experimental apparatus used for investigating the raw material vapors mixing effect on index-profile formation. The apparatus consists of a quasi-preform, a model made of a fused silica glass having similar shape and size to a porous preform, a glass synthesizing torch (Fig.4.4) and a two dimensional temperature measurement apparatus (Thermo Viewer). The quasi-preform is equipped with a hot gas heater in it as shown in the figure to control the surface temperature of the quasi-preform.

A vapor-phase stream of $SiCl_4$–$GeCl_4$ (10 mol%) or $SiCl_4$–$TiCl_4$ (4 mol%) was blown from nozzle I into the flame at a rate of 165 cc/min and a pure $SiCl_4$

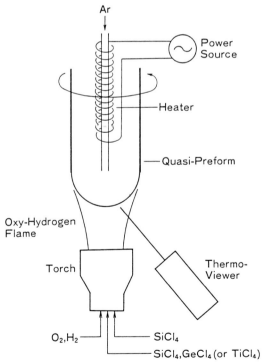

Fig. 4.45. Experimental apparatus used for investigating the mixing effect of raw material vapors on profile formation. [After Sudo[4.57]]

vapor stream was blown from nozzle II at a rate of 0–83 cc/min to induce the mixing effect between nozzle I and nozzle II. The quasi-preform surface temperature is controlled by varying the hot gas heater power in the temperature range from 500°C to 800°C, as measured at its maximum point. Fine oxide particles synthesized by flame hydrolysis reaction were deposited on the quasi-preform surface to form a thin porous glass layer. Each deposition was continued for 1 minute to make a porous glass layer about 1 mm thick, as measured at the bottom surface of the quasi-preform.

Figures 4.46(a) and (b) show measured dopant concentration distribution on the quasi-preform for (a) SiO_2–GeO_2 system and (b) SiO_2–TiO_2 system. Curves A in Fig. 4.46(a) and (b) indicate distributions obtained when (a) $SiCl_4$–$GeCl_4$ or (b) $SiCl_4$–$TiCl_4$ mixture was blown from nozzle I (165 cc/min) and no $SiCl_4$ vapor from nozzle II. Curves B represent the distribution formed when a $SiCl_4$ vapor (83 cc/min) was additionally blown from nozzle II. The difference between curves A and curves B represent the mixing effect of raw materials blown from nozzle I and II.

Even when nozzle II is not used, a graded-like dopant distribution was obtained for the SiO_2–GeO_2 system. This may be attributed to the surface temperature distribution of the quasi-preform such that the temperature was high (650°C) at the center part and decreased gradually along the radial direction of quasi-preform surface. In case of the SiO_2–TiO_2 system with nozzle II not used (curve A in Fig. 4.46(b)), an almost step-like dopant distribution was formed as expected from the experimental results of the previous Section 4.6.3; a small deviation from the ideal step might be attributed rather to nonuniformity of hydrolysis reactions in the radial direction of the flame.

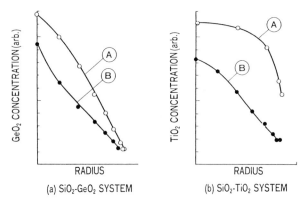

Fig. 4.46. Measured dopant concentration distribution in the quasi-preform. [After Sudo[4.57]]

The graded TiO$_2$ distribution in curve B in Fig. 4.46(b) is attributed largely to the mixing effect of the raw materials, while the graded GeO$_2$ distribution in curve B in Fig. 4.46(a) is caused by the combination of the temperature effect and the mixing effect.

The data for Fig. 4.46, being combined with the data for Fig. 4.44, offer useful principles for precise control over the index profile in an actual VAD process. In the SiO$_2$–GeO$_2$ system, both the flame temperature and the surface temperature distribution on the porous preform must be controlled to make a desired GeO$_2$ distribution profile; the mixing effect of the raw materials blown from multiple nozzles is also useful to modify the index profile. In the case of the SiO$_2$–TiO$_2$ system, the mixing effect of the raw materials is rather dominant in the profile formation, though low loss could not be expected in the system. In the following, the glass composition is restricted to the SiO$_2$–GeO$_2$ system being of practical importance.

4.6.4 Profile control techniques
A) Application of surface temperature effect[54]

On the basis of the profile formation mechanism investigation mentioned in Section 4.6.3, the surface temperature effect on the actual profile formation will be presented. The surface temperature distribution on the porous preform end was formed by regulating the rate of hydrogen gas flow for combustion. The mixture of SiCl$_4$, GeCl$_4$ and POCl$_3$ vapors and the SiCl$_4$ vapor were blown from nozzles I and II, respectively. A typical surface temperature distribution is shown in Fig. 4.43. The temperature distribution along the center line of the preform is indicated on the left in the figure, from which it can be seen that maximum temperature is about 650°C at the center part and the temperature decrease to 300°C gradually along the radial direction.

Interference microscopic photographs of fibers fabricated under (a) 630°C, (b) 680°C and (c) 730°C maximum temperature and (a) 1200°C, (b) 1300°C and (c) 1400°C flame temperature are shown in Fig. 4.47. Profile parameter, α,* for case (a), (b) and (c) were 1.63, 1.95 and 3.94, respectively.

The profile parameter becomes larger with increasing surface temperature, which is due to the substrate temperature dependence of the GeO$_2$ concentration. Figure 4.48 shows the relation between the maximum surface temperature and profile parameter. The relation shown in the figure can be

---

*Profile parameter of graded index fiber is defined by the following equation of refractive-index profile.

$$n^2(r) = \begin{cases} n_0^2 \left\{ 1 - 2\Delta\left(\dfrac{r}{a}\right)^\alpha \right\} & 0 \leq r \leq a \text{ (core region)} \\ n_0^2 (1 - 2\Delta) & a \leq r \leq b \text{ (cladding region)} \end{cases}$$

where n_0 is refractive-index at the core center, Δ is refractive-index difference between core center and cladding, a is core diameter, b is cladding diameter.

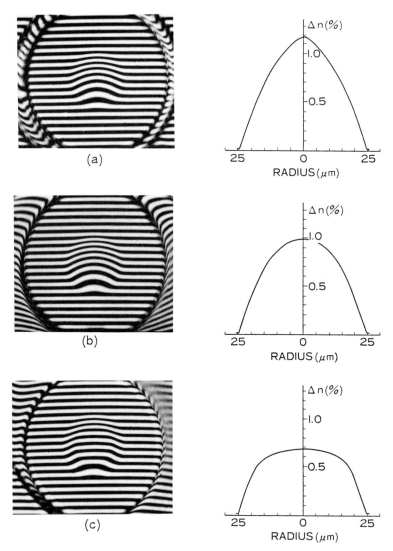

Fig. 4.47. Interference microscope photographs and the refractive-index profiles made under the maximum temperatures of (a) 630°C, (b) 680°C, and (c) 730°C. [After Sudo[4.57]]

reasonably explained using the contour line map of the GeO_2 concentration indicated in Fig. 4.42. When the flame temperature is set at 1200°C and the surface temperature on the porous preform ranges from 630°C to 280°C, which is shown by two-dots-dash-line in Fig. 4.44, the GeO_2 concentration at

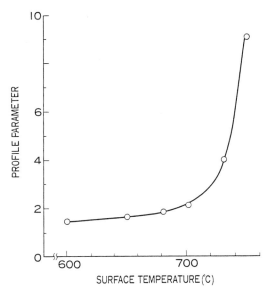

Fig. 4.48. Relation between the maximum surface temperature and the profile parameter. [After Sudo[4.57]]

the center of the porous preform may become about 12 mol% and then decreases gradually in the radial direction.

The GeO_2 concentration in the porous preform fabricated with 680°C maximum surface temperature and 1300°C flame temperature has a maximum of about 10 mol% at the center and the refractive-index profile becomes nearly parabolic (Fig. 4.47(b)), which can be deduced from the dash-line on Fig. 4.44.

When the flame temperature is elevated to 1400°C, the maximum GeO_2 concentration further decreases to about 7 mol% and the index profile becomes step-like (Fig. 4.47(c)), which can be deduced from the dot-dash-line in Fig. 4.44. The result shown in Fig. 4.48 suggests that the refractive-index profile parameter can be controlled in the range of 1 to 10 by varying the porous preform surface temperature.

B) *Application of gas mixing effect*

As seen in the previous section, the refractive-index profile in the SiO_2-GeO_2 system is basically formed by the substrate temperature dependence of the GeO_2 concentration, but the precision of the profile determined by the temperature effect is not very accurate, especially in the peripheral area. In the present section, the gas mixing effect on the actual profile formation is investigated.

The flow rate ratio, R, of $SiCl_4$ vapor blown from nozzle II to $SiCl_4$–$GeCl_4$–$POCl_3$ vapor mixture from nozzle I was varied in three levels of (A) $R=0.1$, (B) $R=0.3$ and (C) $R=0.5$. Figure 4.49 shows the refractive-index profiles for fibers made under the above three flow rate conditions. In each deposition, the maximum temperature was kept at 650°C. The profile parameter for each fiber was evaluated as 2.40 for fiber (A), 1.94 for fiber (B) and 1.83 for fiber (C).

Figure 4.50 shows the relation between the profile parameters and the material gas flow rate ratio. Solid circles and open circles in the figure indicate the profile parameters measured on the fibers fabricated under the maximum temperature conditions of 680°C and 650°C, respectively. It can be seen that the profile parameters decrease monotonically with increasing flow rate ratio,

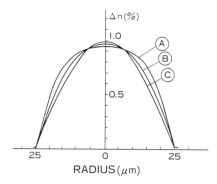

Fig. 4.49. Refractive-index profiles of the fibers made under three flow ratio conditions of (A) $R=0.1$, (B) $R=0.3$, and (C) $R=0.5$. [After Sudo[4.57]]

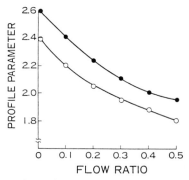

Fig. 4.50. Relation between the gas flow rate ratio, R, and profile parameter. [After Sudo[4.57]]

R. It can also be seen that the parameter at a fixed ratio shows an increase, according to an increase in the maximum temperature.

Figure 4.50 shows that profile parameters for fibers can be controlled to the desired value around 2.0, using the gas mixing effect on the profile formation along with the temperature effect. Figure 4.51 shows the interference microscopic photograph for a typical VAD graded-index fiber.

4.6.5 Fluorine doping

Although fully fluorinated silicon, SiF_4, is a vapor at room temperature, partially fluorinated silica is used for making glass with a lower refractive-index than that of silica and for the termination of dangling bonds in silica glass to reduce the effects of hydrogen permeation.

In the VAD process two methods have been developed to dope fluorine in silica glass.[55,56] One consists of making glass by the hydrolysis of the mixed gas of $SiCl_4$ and SF_6 or CF_4. The other method is based on the thermal treatment of porous preforms in an atmosphere containing SF_6, CF_4 or $CCl_2F-CClF_2$ at about 1000°C for 1 hour. The porous preform is then consolidated into a transparent glass. With the latter technique, it is reported that the refractive-index of silica glass can be reduced to about 0.3 % from the original index[56] although the fluorine doping level depends on the soot density of porous preform as well as on the fluorine concentration in the treating atmosphere. Figure 4.52 shows the relation between soot density and refractive index. SF_6 concentration in the atmosphere was fixed at 4 mol% throughout these experiments. This result also suggests that the refractive-index of the silica glass obtained can be controlled to a desired value by the soot density.

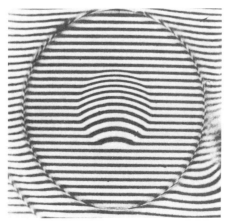

Fig. 4.51. Interference microscopic photograph for a typical VAD graded-index fiber. [After Horiguchi[6.16]]

132   Chapter 4

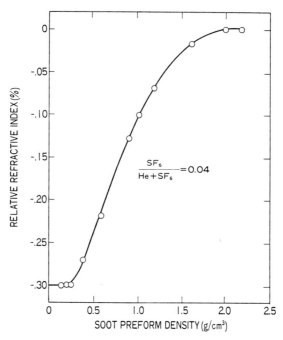

Fig. 4.52. Relation between soot density and refractive-index in fluorine doped fiber. [After T. Miya, Newly developed fluorine doping techniques for VAD process, *Proc. ECOC*, 84, p. 294 (1984)]

REFERENCES

1) T. Izawa, F. Hanawa, S. Kobayashi, S. Shibata, S. Sudo and T. Miyashita, "Optical fibers fabricated by the Vapor-Phase Verneuil method", (in Japanese)presented at the IECE Jpn. Nat. Conf., Tech. Dig., p. 792 (1977).
2) T. Izawa, S. Kobayashi, S. Sudo and F. Hanawa, "Continuous fabrication of highsilica fiber preform", presented at the IOOC'77, Tokyo, Tech. Dig., C1-1 (1977).
3) T. Izawa, T. Miyashita and F. Hanawa, "Continuous optical fiber preform fabrication method", U.S. Patent 4,062,665 (Filed Apr. 5, 1977).
4) T. Izawa, S. Sudo, F. Hanawa and S. Kobayashi, "Continuous fabrication process for optical fiber preforms" (in Japanese), *ECL Tech. J.*, Vol. 26, No. 9, p. 2531 (1979).
5) T. Izawa, S. Sudo and F. Hanawa, "Continuous fabrication process for high silica fiber preforms", *Trans. IECE Jpn.*, Vol. E62, No. 11, p. 779 (1979).
6) S. Sudo, M. Kawachi, T. Edahiro, M. Nakahara, F. Hanawa and S. Tomaru, "Fabrication method of optical fiber preforms", U.K. Patent Appl. GB 2,059,944 (Filed Oct. 1980).
7) S. Sudo, F. Hanawa, M. Kawachi and T. Edahiro, "Torch for synthesizing fine glass particles", Japanese Patent 1,121,387 (1982).
8) S. Sudo, M. Kawachi and F. Hanawa, "Torch for synthesizing fine glass particles", Japanese Patent 1,103,644 (1981).

9) M. Kawachi, S. Tomaru, S. Sudo and T. Edahiro, "Fabrication method of single-mode optical fiber preform", U.S. Patent 4,345,928 (Filed Sept. 19, 1980).
10) K. Chida, F. Hanawa, M. Nakahara and N. Inagaki, "Simultanious dehydration with consolidation for VAD method", *Electron. Lett.*, Vol. 15, No. 25, p. 835 (1979).
11) F. Hanawa, S. Sudo, K. Chida, H. Okazaki and M. Nakahara, "Large preform fabrication using a continuous consolidation technique" (in Japanese), presented at the IECE Jpn. Nat. Conf., Tech. Dig., p. 942 (1981).
12) M. Nakahara, N. Inagaki, K. Yoshida, M. Yoshida and O. Fukuda, "Fabrication of 100 km graded-index fiber from a continuously consolidated VAD preform", Tech. Dig. IOOC'84, p. 100 (1984).
13) S. Sudo, F. Hanawa and T. Edahiro, "Vapor-phase axial deposition method for optical fiber preform fabrication I" (in Japanese), Private Communication Paper in ECL, No. 13128 (1978).
14) S. Sudo, M. Kawachi and F. Hanawa, "Vapor-phase axial deposition method for optical fiber preform fabrication IV" (in Japanese), Private Cmmunication Paper in ECL, No. 14176 (1979).
15) W. G. French and L. J. Pase, "Chemical kinetics of the modified chemical vapor deposition process", presented at he IOOC'77, Tokyo, Tech. Dig., c1-1 (1977).
16) J. W. Hastie, "Molecular basis of flame inhibition", *J. Res. Natl. Bur. Standards—A. Phys. Chem.*, Vol. 77A, No. 6, p. 733 (1973).
17) M. J. Day, G. Dixon-Lewis and K. Thompson, "Flame structure and flame reaction kinetics VI", *Proc. R. Soc. Lond.*, A330, p. 199 (1972).
18) M. Kawachi, A. Kawana and T. Miyashita, "Low-loss single mode fiber at the material dispersion-free wavelength of 1.27 μm", *Electron. Lett.*, Vol. 13, p. 442 (1977).
19) H. Suda, S. Sudo and M. Nakahara, "Deposition mechanism of fine glass particles and high-rate production of fiber preforms in the VAD process", *Electron. Lett.*, Vol. 18, No. 18, p. 779 (1982).
20) H. Suda, S. Sudo and M. Nakahara, "Fine glass particle-deposition mechanism in the VAD process", *Fiber and Integrated Optics*, Vol. 4, No. 4, p. 427 (1983).
21) J. Frenkel, "Viscous flow of crystalline bodies under the action of surface tension", *J. Phys. (USSR)*, Vol. 9, No. 5, p. 385 (1945).
22) G. W. Shere, "Sintering of low-density glass I", *J. Am. Ceram. Soc.*, Vol. 61, Nos. 5-6, p. 236 (1977).
23) S. Sudo, M. Kawachi and T. Edahiro, "Analysis for consolidation process of porous glass bodies" (in Japanese), presented at Nat. Conv. Jpn. Soc. Appl. Phys., Tech. Dig., 1p-B-5 (Sept. 1979).
24) S. Sudo, T. Edahiro and M. Kawachi, "Sintering process of porous preforms made by a VAD method for optical fiber fabrication", *Trans. IECE Jpn.*, Vol. E63, No. 10, p. 731 (1980).
25) W. G. Perkins and D. R. Begeal, "Diffusion and permeation of He, Ne, Ar, Kr and D_2 through silicon oxide thin film", *J. Chem. Phys.*, Vol. 54, No. 4, p. 1683 (1971).
26) A. Petzold, "Versuch zur klassifizierung von komponenten nach ihrem einfluss auf die oberfachenspannung vcon silikatschmelzen", Silikattechnik Jg. Helf 1, p. 11 (1954).
27) S. Sudo, M. Kawachi, N. Shibata and t. Edahiro, "PH-ion behavior and its reduction in porous preform for optical fiber fabrication" (in Japanese), presented at Nat. Conv. Jpn. Soc. Appl. Phys., Tech. Dig., 3p-p-7 (Oct. 1978).
28) H. L. Hair, "Hydroxyl group on silica surface", *J. Non-Cryst. Solids*, Vol. 19, p. 299 (1975).
29) J. B. Peri, "Infrared study of OH and NH_3 groups on the surface of a dry silica aerogel", *J. Phys. Chem.*, Vol. 70, No. 9, p. 2937 (1966).
30) L. H. Little (translated by Hasegawa *et al.*), "*Adhesion and Infrared Absorption Spectra*", Kagaku Doginsha (1975).

31) S. Sudo, M. Kawachi, T. Edahiro and N. Inagaki, "Dehydration and consolidation technique in the Vapor-Phase Axial Deposition method" (in Japanese), *ECL Tech. J.*, Vol. 29, No. 10, p. 1719 (1980).
32) T. Edahiro, M. Kawachi, S. Sudo and H. Takata, "OH-ion reduction in VAD fibers", *Electron. Lett.*, Vol. 15, No. 16, p. 482 (1979).
33) S. Sudo, M. Kawachi, T. Izawa, T. Edahiro, T. Shioda and H. Gotoh, "Low OH content optical fiber fabricated by Vapor-Phase Axial Deposition method", *Electron. Lett.*, Vol. 17, p. 534 (1978).
34) S. Sudo, M. Kawachi, N. Shibata and T. Edahiro, "Transmission loss characteristics of OH-content optical fibers fabricated by the Vapor-Phase Axial Deposition method", presented at Nat. Conf. IECE Jpn., Tech. Dig., 378 (Sept. 1978).
35) T. Edahiro, M. Kawachi, S. Sudo and N. Inagaki, "OH-ion reduction in the optical fibers fabricated by the Vapor-Phase Axial Deposition method", *Trans. IECE Jpn.*, Vol. E63, No. 8, p. 574 (1980).
36) J. F. Shackelford and J. S. Masaryk, "The thermodynamics of water and hydrogen solubility in fused silica", *J. Non-Cryst. Solids*, Vol. 21, p. 55 (1976).
37) N. Uchida, N. Uesugi, Y. Murakami, M. Nakahara, T. Tanifuji and N. Inagaki, "Infrared loss increase in silica optical fiber due to chemical reductin of hydrogen", Tech. Dig. of ECOC'83, post dead line paper (1983).
38) F. Hanawa, S. Sudo, M. Kawachi and M. Nakahara, "Fabrication of completely OH-free VAD fiber", *Electron. Lett.*, Vol. 16, No. 18, p. 699 (1981).
39) S. Sudo, S. Kabayashi, T. Izawa and Y. Masuda, "Fabrication of graded-index fiber preforms by the Vapor-Phase Veruneuil method" (in Japanese), presented at Nat. Conf. IECE Jpn. Tech. Dig., 795 (Mar. 1977).
40) K. Sanada, T. Shioda, T. Moriyama, K. Inada, M. Kawachi and H. Takata, "Refractive-index profile control of the Vapor-Phase Axial Deposition", *Proc. Opt. Commun. Conf.*, 5.1-1 (1980).
41) M. Kawachi, S. Sudo, N. Shibata and T. Edahiro, "Deposition properties of SiO_2–GeO_2 particles in the flame hydrolysis reaction for optical fiber fabrication", *Jpn. J. Appl. Phys.*, Vol. 19, No. 2, p. 169 (1980).
42) M. Kawachi, S. Sudo, N. Shibata, S. Tomaru and T. Edahiro, "Structure of fine glass particles in the flame hydrolysis reaction for optical fiber fabrication" (in Japanese), presented at Nat. Conf. IECE Jpn., Tech. Dig., 941 (Mar. 1980).
43) T. Edahiro M. Kawachi, S. Sudo and S. Tomaru, "Deposition properties of high silica particles in the flame hydrolysis reaction for optical fiber fabrication", *Jpn. J. Appl. Phys.*, Vol. 19, No. 11, p. 2047 (1980).
44) N. Borreli, "The infrared spectra of SiO_2–GeO_2 glass", *Phys. Chem. Glasses*, Vol. 10, p. 43 (1969).
45) A. S. Tenny and J. Wong, *J. Chem. Phys.*, Vol. 56, p. 5516 (1972).
46) C. F. Smith, R. A. Condrate and W. E. Votava, *Appl. Spectroscopy*, Vol. 29, p. 79 (1975).
47) ASTM cards.
48) G. V. Samsov, "*The Oxide Handbook*", p. 178, IFI/Plenum, New York-Washington-London (1973).
49) M. Kawachi, S. Sudo and T. Edahiro, "Threshold gas flow rate of halide material for the formation of oxide particles in the VAD process for optical fiber fabrication", *Jpn. J. Appl. Phys.*, Vol. 20, No. 4, p. 709 (1981).
50) CRC Handbook of chemistry and physics, 52nd ed., p. B-93, CRC Press, Cleaveland (1973).
51) S. Sudo, H. Suda, M. Kawachi and M. Nakahara, "Formation mechanism analysis of SiO_2–GeO_2 fine particles for optical fiber fabrication" (in Japanese), presented at Nat. Conf. Jpn. Soc. Appl. Phys., Tech. Dig., 31p-D-13 (May 1981).
52) S. Sudo, H. Suda, M. Nakahara and Y. Maeda, "Formation mechanism analysis of

SiO_2–GeO_2 fine particles for optical fiber fabrication (II)" (in Japanese), presented at Nat. Conf. Jpn. Soc. Appl. Phys., Tech. Dig., 9p-S-5 (Sept. 1981).
53) M. Kawachi, S. Sudo and T. Edahiro, "Formation mechanism of refractive-index profile in the VAD method", *Trans. IECE Jpn.*, Vol. C65, No. 4 (1982).
54) S. Sudo, M. Kawachi, H. Suda, M. Nakahara and T. Edahiro, "Refractive-index profile control techniques in the Vapor-Phase Axial Deposition method", *Trans. IECE Jpn.*, Vol. E64, No. 8, p. 536 (1981).
55) S. Sudo, T. Miya and M. Nakahara, "Fabrication of fluorine doping VAD preforms" (in Japanese), presented at Nat. Conf. IECE Jpn., Tech. Dig., 1095 (1983).
56) T. Miya, M. Nakahara, F. Hanawa and Y. Ohmori, "Newly developed fluorine doping techniques for VAD process", *Tech. Dig. ECOC* '84, p. 294 (1984).
57) S. Sudo, "Studies on the Vapor-Phase Axial Deposition Method for optical fiber fabrication", Ph. D. Thesis, Tokyo Univ. (1982).

Chapter 5

FABRICATION PROCESSES OF MULTI-COMPONENT GLASS FIBERS AND NON-SILICA FIBERS

Fabrication processes of various kinds of fibers except for high-silica fibers will be mentioned briefly to make it easy to understand their fiber performances. Major differences between high silica fiber and other kinds of fibers lie in both processes of raw material purification and fiber drawing processes. As for most fibers, their performances can be estimated from the material characteristics with reasonably good accuracy. As for crystalline fibers made of alkali-halides, however, there is a big difference between the estimated loss and actually obtained fiber loss. This may be understood from the drawing processes, which is being used today. Transmission characteristics of these fibers will be described in Chapter 6.

5.1 Multi-Component Glass Fibers

5.1.1 Fabrication process

Although the optical loss of multi-component glasses has been reduced to 3.4 dB/km at 0.84 μm wavelength in 1977,[1] it has not been popular to use them for the optical communication systems. The essential problems that need to be solved are derived from the glass composition containing a large amount of network modifiers such as alkali metal oxides and alkali-earth metal oxides. The purification of raw materials for alkali metal oxides and alkali-earth metal oxides is not easy, resultingly it is very difficult to fabricate low-loss fibers, less than 1 dB/km, with multi-component glasses. Furthermore the chemical durability and mechanical strength of multi-component glass fibers are much poorer compared to that of high silica glass fibers.[2] These, above all, could be attributed to the network modifiers, that is, to alkali metal oxides. Though multi-component glasses have all these disadvantages, they are still used for special purposes such as high-numerical-aperture fibers for image guides and short-length signal transmission, because many kinds of glasses with large variety of optical and thermal properties can be easily prepared at comparatively low cost. An example of multi-component glass fiber fabrication will be mentioned briefly next.

A typical glass composition for optical fibers is soda-lime-silica because of its stable glassy state, relatively low dispersion and low Rayleigh scattering loss.[3] In this example, SiO_2, GeO_2, Na_2O, CaO, Li_2O, and MgO were used as glass components. The following starting materials were used: ultra-high purity SiO_2 prepared by flame hydrolysis of SiH_4, distilled $Ge(OC_4H_9)_4$ for GeO_2, $NaNO_3$, $LiNO_3$, $Ca(NO_3)_2$ and $MgSO_4$.[4] These nitrates and sulfate were purified by ion exchange process or solvent extraction process.

A block diagram for the preparation of high purity glass is shown in Fig. 5.1. In order to minimize contamination during preparation process, purification, condensation and mixing of the starting materials were carried out in a closed system. This system was set up using plastic tubes and connectors among each unit. Purified raw materials were carried with water into a vacuum rotary evaporator, where the starting materials were mixed and condensed to slurry. Then, this slurry was transported into a fused silica crusible by a plastic-pump and completely dried in a resistance furnace. Melting and refining of glasses were carried out at 1400°C for 1 hour under clean and dry N_2–O_2 gas flow.

An illustration showing fiber drawing is given in Fig. 5.2. Glass blocks made by above process were cut into several pieces, polished with abrasives, cleaned with a supersonic-wave cleaner in HNO_3–H_2SO_4 solution and rinsed in distilled water. The core and clad glasses were put into a plutinum double crucible and heated up to 750°C to draw into fibers. The fiber drawing was carried out in a cylindrical fused silica cover under clean and dried N_2 gas. After drawing, the glass fibers were coated with silicone.

5.1.2 Purification of raw materials

Raw materials of above mentioned multi-component glass are SiO_2, $Ge(OC_4H_9)_4$, $NaNO_3$, $LiNO_3$, $Ca(NO_3)_2$ and $MgSO_4$. Alkali nitrates, alkali-earth metal nitrate and alkali-earth metal sulfate are purified by ion exchange or solvent extraction methods. The purification processes use the chemical reactions of chelates.

Chelate resin having a similar structure to ethylene-diamine-tetraacetic-acid (EDTA) was used as ion exchange resin. Metal ions react with EDTA and make chelate compounds as shown in Fig. 5.3. The chelate formation reaction between metal ion, M^{n+}, and chelate reagent, HnZ, expressed in eq. (5.1), balances at K_{MZ}.

$$M^{m+} + Z^{n-} \rightleftharpoons MZ^{m-n} \tag{5.1}$$

where

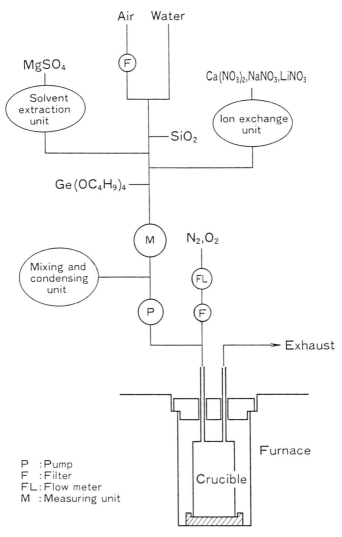

Fig. 5.1. Block diagram of multicomponent glass preparation system. [After Takahashi[(5.4)]]

$$K_{MZ} = \frac{[MZ^{m-n}]}{[M^{m+}][Z^{n-}]} . \qquad (5.2)$$

Metal ion exchange reaction with chelate, $M_2Z^{m_2-n}$, and impurity ion, M_1, shown in eq. (5.3) balances at K_{M_1,M_2}.

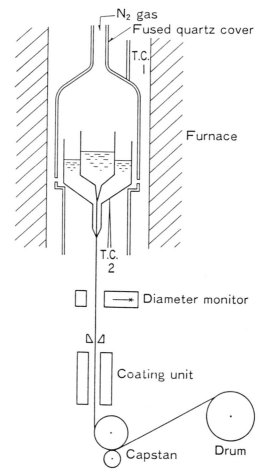

Fig. 5.2. Schematic diagram of multicomponent glass fiber drawing system. [After Takahashi[5.4]]

$$M_2Z^{m_2-n} + M_1^{m_1+} \rightleftharpoons M_1Z^{m_1-n} + M_2^{m_2+} \qquad (5.3)$$

where

$$K_{M^1,M^2} = \frac{[M_1Z^{m_1-n}][M_2^{m_2+}]}{[M_2Z^{m_2-n}][M_1^{m_1+}]}. \qquad (5.4)$$

Purification of alkali metal or alkali-earth metal ions (supposed to be M_2) using the reaction of eq. (5.3) can be realized when K_{M_1,M_2} is larger than one.[5] When

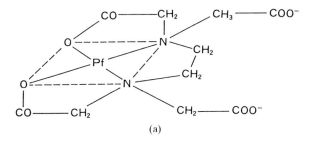

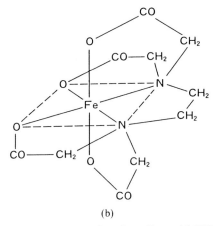

Fig. 5.3. Chelate compounds made by reaction of metal ions with EDTA. (a) 4-sites ligand, (b) 6-sites ligand. [After Takahashi[5.25)]]

log K_{M_1,M_2} is larger than 9, M_1, the ion concentration will be less than 1 ppb. Figure 5.4 shows the values of log K_{M_1,M_2} for various alkali metal and alkali-earth metal ions. When these values are larger than 8, the nitrates and sulfates could be purified to less than 0.1 ppm. It is understood from the figure that the purification of alkali-earth metal ions such as Ba^{2+}, Mg^{2+} and Ca^{2+} is not easy only using these chelate reactions.

The solution extraction method also uses chelate reactions. The process using APDC (ammonium-pyrrolidine-dithiocarbamate) and MIBK (methyl-isobutyl-ketone) is shown in Fig. 5.5.[6)]

5.1.3 Bubble formation

Bubble formation in the double crucible process for fiber drawing, which can not be seen in the preform drawing method, is an important factor to be solved in order to make low-loss fibers. There is a difference in alkali-ion

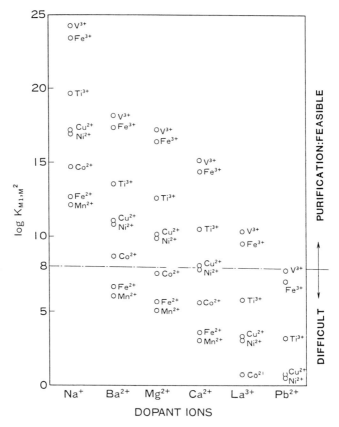

Fig. 5.4. The difference of chelate formation constant between transition metal ions and dopant ions. [After Takahashi[5.25]]

concentration between the core and cladding glasses. They form a battery with platinum as is shown in Fig. 5.6.[7] At the electrode immersed in the higher ion concentration glass, O^{2-} becomes O_2 and electrons as is shown in the following equations.[7] At the second electrode immersed in the lower ion concentration glass, oxygen gas in the atmosphere is taken up in the molten glass and becomes O^{2-}.

$$\text{Electrode I:} \quad 2O^{2-} \longrightarrow O_2 + 4e^- \quad \text{(bubble formation)} \tag{5.5}$$

$$\text{Electrode II:} \quad 4e^- + O_2 \longrightarrow 2O^{2-}. \tag{5.6}$$

These bubbles are formed on the inner crucible surface, when the alkali-ion

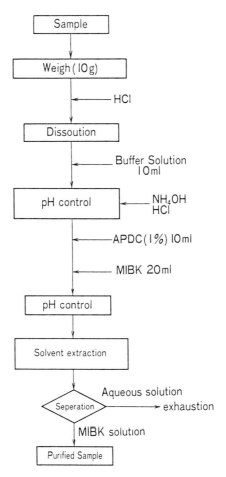

Fig. 5.5. APDC-MIBK solvent extraction process for high-purity multi-component glass. [After Takahashi[5.25)]]

concentration of core glass is higher than that of cladding glass. Some of the bubbles are introduced into the core-cladding interface of the fiber obtained as shown in Fig. 5.7. Therefore, it is preferable to have the same alkali-ion concentration in the core and cladding glasses and to be drawn in an oxygen-free atmosphere.

5.2 Fluoride Glass Fibers

Fluoride glass is transparent in the infrared regions where silica glass has comparatively high transmission losses. There are many kinds of fluoride

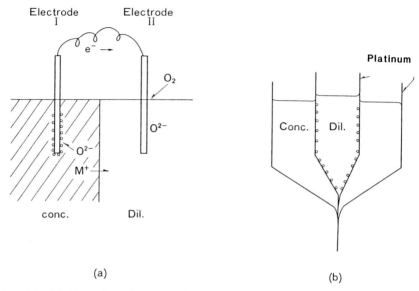

Fig. 5.6. (a) Formation of battery with the concentration difference of alkali ions in multicomponent glass, (b) Bubble formation in a double crucible. [After Takahashi[5.25]]

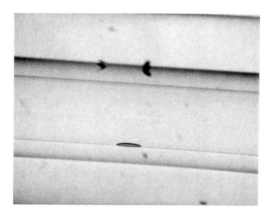

Fig. 5.7. Bubble formed with battery effect at the boundary layer between core and cladding of multi-component glass fiber. [After Takahashi[5.25]]

glass systems, and the systems are devided into three groups; BeF_4-based glasses,[8] AlF_3-based glasses[9] and ZrF_4-based glasses.[10] These fluoride glasses are much more unstable compared to oxide glass systems. Some fluoride systems, however, can be fully vitrified. Even if a stable fluoride glass can be made, it sometimes devitrifies or crystalizes during the fiber drawing process. Therefore, it is important to select a very stable glass system for fiber fabrication.

A typical fluoride glass is ZrF_4–BaF_2–GdF_3.[11,12] This system is comparatively stable. The stable glass constituent of this system is shown in Fig. 5.8 by encircled region with dotted line. The stable fiber-producible system is shown by encircled region with solid line, which is narrower than the glass forming region. Figure 5.9 shows the refractive indices of the glasses.[12] Figure 5.10 shows a process[11] for fiber preparation. Starting materials were ZrF_4, GdF_3, AlF_3 and SbF_3 of commercially available purity (99.99%). ZrF_4, AlF_3 and SbF_3 were purified by sublimation under a low pressure of 2–3 mmHg at 850°C for ZrF_4, 1000°C for AlF_3 and 300°C for SbF_3, respectively using a sublimation apparatus as shown in Fig. 5.11.[12] BaF_2 and GdF_3 were purified by sublimating out other transition metal fluorides from the raw materials. These temperatures were determined from their vapor pressure as shown in Fig. 5.12.[12]

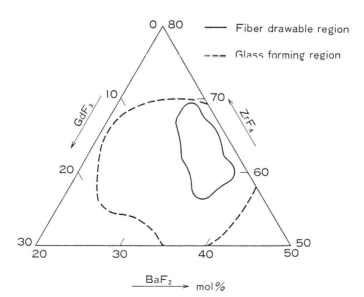

Fig. 5.8. Fiber drawable region and glass forming region for BaF_2-GdF_3-ZrF_4 glass. [After Mitachi et al.[5,11]]

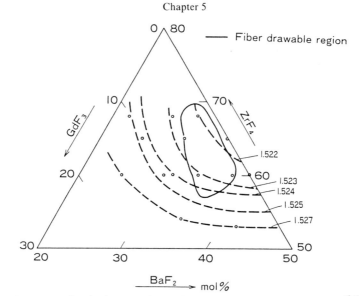

Fig. 5.9. Refractive-indexes of BaF_2-GdF_3-ZrF_4 glass. [After Mitachi et al.[5.11]]

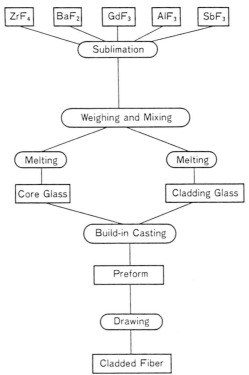

Fig. 5.10. Preparation process for fluoride glass fibers. [After Mitachi et al.[5.12]]

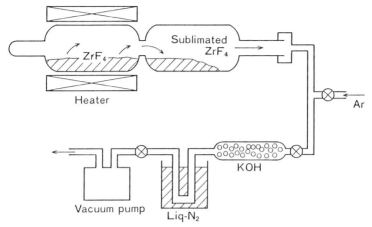

Fig. 5.11. Sublimation system for the purification of fluorides. [After Mitachi et al.[5.11]]

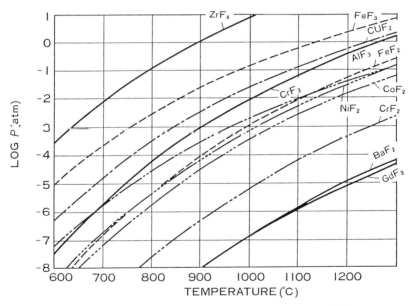

Fig. 5.12. Vapor pressure of fluorides. [After Mitachi et al.[5.11]]

Fabrication process of a low-loss fluoride fiber is as follows: The core glass composition was $30.4BaF_2$–$3.7GdF_3$–$58.1ZrF_4$–$3.8AlF_3$–$4.0SbF_3$ and the cladding glass composition was $29.8BaF_2$–$3.6GdF_3$–$56.9ZrF_4$–$5.7AlF_3$–$4.0SbF_3$. AlF_3 and SbF_3 were added to control the refractive-index and to stabilize the glassy state. These raw materials were mixed and treated at 400°C for 30 min with NH_4F-HF to fluorize perfectly. Then, it was melted at 900°C for 2 hours in a gold crucible. Fiber preforms were then made by a "built-in casting" process as shown schematically in Fig. 5.13. The molten cladding glass was poured into a cylindrical brass mold preheated to around the glass transition temperature. This was immediately turned over so that the melt in the central part of the mold ran out, and the core glass melt was poured into the central cylindrical hollow and then cooled down. The preform obtained was jacketed with a plastic tube and heated by an electric furnace at a temperature around 350–400°C to draw into a fiber, with precise control of drawing temperature to avoid crystallization and resultant scattering-loss increase.

5.3 Chalcogenide Glass Fibers

The raw materials of chalcogenide glasses for infrared optical fiber use are S, Se, As and Ge.[13-15] Commercially available Se and S are not sufficiently pure. These materials are usually purified by distillation in argon gas atmosphere, or S_2Cl_2 atmosphere in case of S, in a glass tube in order to remove OH, SH, SeH and carbon.[15] The distillation apparatus is similar to that used for the purification of fluorides mentioned in Section 5.2. Ge and As are available with high purity better than six-9. Therefore, these materials can be used only by etching off the surface contamination. Reagents of glass are

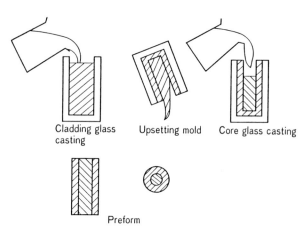

Cladding glass casting   Upsetting mold   Core glass casting

Preform

Fig. 5.13. Preform preparation process of fluoride glass fiber by build-in casting method. [After Mitachi et al.[5.11]]

weighed out into a quartz ampoule, which is then evacuated and sealed still under a vacuum. The sealed ampoules are heated at 700–900°C for 20–100 hours in a rocking furnace to mix the constituents and the raw materials are made into a chalcogenide glass. The ampoules are withdrawn from the furnace to cool naturally in air. Glass rods 8–15 mm in diameter and 80–120 mm in length are obtained. Then, the rods are drawn into unclad fibers in Ar atmosphere. The unclad fibers are used with a loose fitting polyethylene tube. It is not easy to make clad fiber with rod-in-tube method, because of the difficulties in fabricating chalcogenide glass tube with optical polished surface. Usually the double crucible method is used for making clad fibers. The crucible is made of Pyrex glass and the design is similar to the conventional platinum double crucible used for making multi-component silicate glass fibers.

It is noted that chalcogenide glasses have the tendency to devitrify during the fiber drawing process. As_xS_{100-x} becomes glassy when x is smaller than 44, however, devitrification in the rod drawing process is observed when x is larger than 42, and in the crucible drawing process, it is observed when x is larger than 39.[15] Transmission loss of the clad fibers fabricated by the double crucible method is higher than that of the unclad fibers made by rod drawing. It is estimated to be due to the nucleation of crystals. Figures 5.14 and 5.15 show glass forming and fiber drawing composition for the As–Ge–Se and

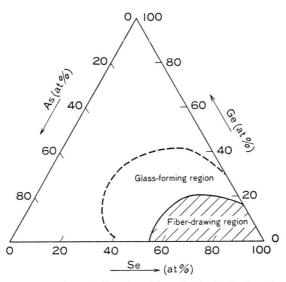

Fig. 5.14. Glass forming region and fiber drawable region in As-Ge-Se system chalcogenide glass. [After Kanamori et al.[1.34]]

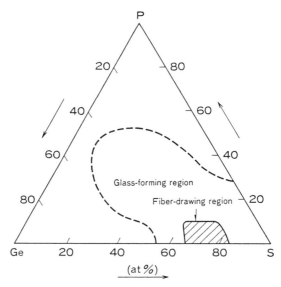

Fig. 5.15. Glass forming region and fiber drawable region in Ge-P-S system chalcogenide glass. [After Kanamori et al.[1.34]]

Ge-P-S glass systems.[15] Then hatched regions in the figures are sufficiently stable to make fibers.

5.4 Crystalline Fibers

Polycrystalline fibers are made by a hot extrusion technique. The most widely studied polycrystalline fibers are KRS-5 (thallium bromoide)[16] and AgCl[17] fibers. The extrusion temperatures are generally between 200°C and 350°C and the extrusion force may be as high as 40 kN for rods of 15 mm in diameter. Fibers have been extruded with diameters ranging from 75 μm to 1 mm and with lengths as long as 100 m. KRS-5 fibers are used with loose fitting polyethylene tubing.[16]

Single-crystalline fibers are made by EFG (edge-defined film-fed growth) method with AgBr[18,19] and CsI.[20] Molten halide is contained in a fused-quartz U tube, which is kept in an oven. Single crystal fiber growth occurs close to the exit of a nozzle, which terminates one arm of the U tube. The growth rate is controlled by N_2 gas pressure applied to the other arm of the U tube. A small oven around the nozzle-tip and a water cooled element control the interface between liquid and crystal state precisely. Clear fibers with diameters between 0.35 mm and 0.75 mm have been grown at rates up to 20 mm/min.[18]

The other method used to make single-crystal fibers is the travelling zone method.[21] Two sets of synchronously driven wheels move an extruded polycrystalline fiber through a small heater coil that is used to form a melt zone with a length between 0.1 and 4 times that of the fiber diameter. As the fiber exits the melt zone, it freezes into a single crystal. The growth rates are approximately 10 mm/min.

Single-crystal fibers made by the above mentioned method have rough surface which in typical for this growing process. Therefore, it seems to be difficult to fabricate low-loss fibers with these methods.

5.5 Plastic Fibers

Low-loss plastic fibers are made of poly-methyl methacrylate (PMMA) or polystyrene.[22] PMMA core fiber is made from the monomer. Usually, the monomer includes polymerization inhibitors, and they are eliminated by rinsing with NaOH solution in the first step. After washing residual NaOH away with pure water, the monomer is dried by adding Na_2SO_4 and $Ca(OH)_2$. The dried monomer is distilled under reduced pressure and the middle fraction is collected. The purified MMA is poured into the main distillation vessel and distilled into the polymerization vessel in the fiber fabrication apparatus as shown in Fig. 5.16.[22] Polymerization initiator (0.01 mol/l of azo-*tert*-butane) and chain transfer agent (0.03 mol/l of *n*-butyl mercaptan) are also distilled into the polymerization vessel. After mixing the monomer solution, the vessel is cooled with liquid nitrogen and kept under vacuum to eliminate the dissolved air. The vessel is heated in an electric furnace at 135°C for 12 hours to polymerize the monomer. The temperature is gradually raised to 180°C and kept at that temperature for 12 hours to complete the polymerization.

The temperature of the vessel is raised to a fiber drawing temperature and the melted polymer is pressurised from the upper end of the vessel by dry nitrogen gas. Fibers are drawn out from the nozzle at a diameter controlled by regulating the temperature, gas pressure and drawing velocity. The fibers are cladded immediately with a melted cladding material such as room temperature vulcanized (RTV) silicone or fluoroalkyl methacrylate polymer.

Deuterated PMMA is used for the fabrication of low-loss plastic fibers because it is expected that deuterated PMMA core fibers shift the optical windows to a higher wavelength region.[22,23] Deuterated MMA is synthesized through acetonecyanohydrin method using deuterated acetone as shown in the following equations.[22]

$$\begin{array}{c} CD_3 \\ \diagdown \\ C=O + HCN \\ \diagup \\ CD_3 \end{array} \longrightarrow \begin{array}{c} CD_3 OH \\ \diagdown \diagup \\ C \\ \diagup \diagdown \\ CD_3 CN \end{array}$$

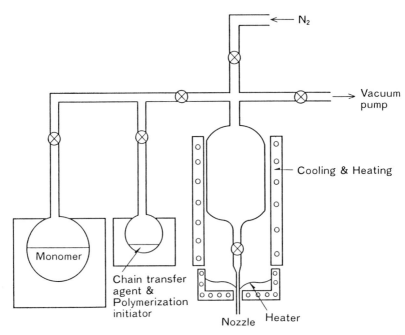

Fig. 5.16. Polymerization and fiber drawing apparatus for plastic fibers. [After Kaino et al.[1.32]]

$$\underset{CD_3}{\overset{CD_3}{>}}C\underset{CN}{\overset{OH}{<}} + H_2SO_4 \longrightarrow \underset{CD_2}{\overset{CD_3}{>}}CONH_2 \cdot H_2SO_4$$

$$\xrightarrow{CD_3OD} \underset{CD_2}{\overset{CD_3}{>}}C\text{-}COOCD_3$$

Polymerization and fiber drawing processes are almost the same as used for PMMA fibers.

Polystyrene fibers are made from purified styrene monomer and 0.04 mol/l of n-butyl mercaptan as a chain transfer agent.[23] The monomer solution is polymerized at 135°C for 16 hours with similar apparatus to the PMMA fiber fabrication. Ethylenevinylacetate copolymer is used as a cladding material.

REFERENCES

1) K. J. Beales, C. R. Day, W. J. Dancan and G. R. Newns, "Low-loss compound-glass optical fiber", *Electron. Lett.*, Vol. 13, p. 755 (1977).
2) S. Shibata, S. Takahashi, S. Mitachi and M. Yasu, "Highly reliable multi-component glasses" (in Japanese), presented at the IECE Japan Nat. Conf., Paper 4-184 (1979).
3) S. Shibata and S. Takahashi, "Effect of some manufacturing conditions on the optical loss of compound glass fibers", *J. Non-Cryst. Solids*, Vol. 23, p. 111 (1977).
4) S. Takahashi, S. Shibata and M. Yasu, "Preparation of low loss multi-component glass fiber", *Rev. ECL*, Vol. 27, p. 123 (1979).
5) K. Ueno, "*Chelate Titration Method*", Nankodo Press, Tokyo, p. 31 (1969).
6) S. Takahashi and M. Yasu, "Analysis of transition metal impurities in optical fiber glasses and their raw materials" (in Japanese), presented at the 30th anual meeting of Japan Chemical Society, paper 3H29 (1974).
7) J. H. Cowan, W. M. Buehl and J. R. Hutchins, "III: An electrochemical theory for oxygen reboil", *J. Am. Ceram. Soc.*, Vol. 49, p. 559 (1966).
8) K. H. Sun, "Fluoride glasses", *Glass Technol.*, Vol. 20, p. 36 (1979).
9) K. H. Sun, "Fluoride glass", U.S. Patent 2,466,509 (1949).
10) M. Poulain, J. Lucas, and P. Brun, "Verres fluores au tetrafluorure de zironium ptoprietes optiques d'un verre dope au Nd^{3+}", *Mater. Res. Bull.*, Vol. 10, p. 243 (1975).
11) S. Mitachi *et al.*, "Fluoride glass fiber for infrared transmission", *Jpn. J. Appl. Phys.*, Vol. 19, p. L313 (1980).
12) S. Mitachi, Y. Ohishi and S. Takahasi, "Prepaiation of fluoride optical fiber", *Rev. ECL*, Vol. 32, p. 461 (1984).
13) N. S. Kapany and R. J. Simms, "Recent development of infrared fiber optics", *Infrared Phys.*, Vol. 5, p. 69 (1965).
14) S. Shibata, Y. Terunuma and T. Manabe, "Ge–P–S chalcogenide glass fibers", *Jpn. J. Appl. Phys.*, Vol. 19, p. L603 (1980).
15) T. Kanamori, Y. Terunuma and T. Miyashita, "Preparation of chalcogenide optical fiber", *Rev. ECL*, Vol. 32, p. 469 (1984).
16) D. A. Pinnow, A. L. Gentile, A. G. Standlee, A. G. Timper and L. M. Hobrock, *Appl. Phys. Lett.*, Vol. 33, p. 28 (1978).
17) J. S. Garfunkel, R. A. Skogman and R. A. Walterson, "Infrared transmitting fibers of polycrystalline silver halides", IEEE/OSA, Laser Engineering and Application, Optical Communication 8.1 (1979).
18) T. J. Bridges, J. S. Hasick and A. R. Strand, "Single-crystal AgBr infrared optical fibers", *Opt. Lett.*, Vol. 5, p. 85 (1980).
19) G. E. Peterson, "Optical waveguide material: the present and the future", Topical Meeting on Optical Fiber Communication, Washington D.C., Tech. Dig., TuA4 (1979).
20) Y. Okamura, Y. Mimura, Y. Komazawa and C. Ota, "CsI crystalline fiber for infrared transmission", *Jpn. J. Appl. Phys.*, Vol. 19, p. L649 (1980).
21) Y. Mimura, Y. Okamura, Y. Komazawa and C. Ota, "Growth of fiber crystals for infrared optical waveguides", *Jpn. J. Appl. Phys.*, Vol. 19, p. L269 (1980).
22) T. Kaino, M. Fujiki and K. Jinguji, "Preparation of plastic optical fibers", *Rev. ECL*, Vol. 32, p. 478 (1984).
23) H. M. Schleinitz, "Ductile plastic optical fibers with improved visible and near infrared transmission", *Int. Wire & Cable Symp.*, Vol. 26, p. 352 (1977).
24) T. Kaino, M. Fujiki and S. Nara, "Low-loss polystyrene core-optical fiber", *J. Appl. Phys.*, Vol. 52, p. 7061 (1981).
25) S. Takahashi, "Studies on the qualification of glass fibers for optical communications", Ph. D. Thesis, Waseda Univ. (1979).

Chapter 6

TRANSMISSION CHARACTERISTICS OF OPTICAL FIBERS

The transmission characteristics of various fibers will be described by the transmission loss characteristics. High silica fibers fabricated by MCVD, OVD and VAD methods have almost the same transmission performance, however, there are small differences in the three fibers mainly in the transmission bandwidth. It comes from the difference in the refractive-index profile such as center dip and index fluctuations. In addition to high-silica fibers, transmission characteristics of multicomponent glass fibers and non-silica fibers such as fluoride fibers, chalcogenide fibers, alkali halide crystal fibers and plastic fibers will be described.

6.1 Single Mode Fibers

6.1.1 Transmission loss

Transmission loss of a single mode fiber is about the same as the estimated loss from various loss origins described in Chapter 2. When the oxyhydrogen ions are reduced to less than 1 ppb, the minimum loss of GeO_2-doped fibers is estimated to be less than 0.18 dB/km.[1] Actually, the transmission loss of OH-free single-mode fibers fabricated by MCVD method and VAD method is less than 0.2 dB/km.[2,3] It has been reported that the minimum loss of a fiber fabricated by VAD method was 0.17 dB/km.[4] The core glass was GeO_2-doped silica and the relative refractive-index difference between the core and the cladding was 0.3%, the fiber length was about 10 km. The Rayleigh scattering coefficient, A, was as small as 0.89. Figure 6.1 shows the loss spectrum of an OH-free single-mode fiber.[4] The OH content was estimated to be 0.02 ppb, which corresponds to the loss of 0.01 dB/km. It has also been reported that a fiber fabricated by MCVD method has a loss of 0.157 ± 0.005 dB/km at 1.57 μm.[5]

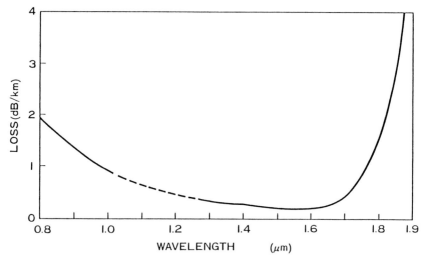

Fig. 6.1. Loss spectrum of ultra-low loss VAD single mode fiber: fiberlength; 10 km, refractive-index difference; 0.3%. [After Kosaka et al.[6.4]]

6.1.2 Transmission bandwidth
A) Characterization technique

There are many kinds of the measurement methods for determining the dispersion characteristics. A typical and simple method is the time domain pulse-delay measurement using a stimulated Raman emission from a GeO_2-doped fiber. The layout diagram of the measurement apparatus is shown in Fig. 6.2(a) schematically.[6] Subnanosecond pulses were obtained by high order sequential stimulated Raman emission in a 1000-m-long GeO_2-doped single-mode fiber, which was excited by Q-switched and mode-locked YAG laser pulses (wavelength: 1.06 μm). Raman laser light, whose spectrum is shown in Fig. 6.2(b), is monochromated and fed into the test fiber. The pulse delay is measured as the time deference between the pulse transmitted through the test fiber and a directly detected pulse.

The delay curves were fitted to the experimental curves using the least square method. The dispersion curves were obtained by calculating the first derivative of the delay curve with respect to wavelength.

B) Dispersion of single mode fibers

The dispersion of single mode fiber is controlled by the material dispersion and the structural dispersion. The total dipersion of a typical single-mode fiber, whose refractive-index difference is around 0.3% and the core radius is about 8 μm, is zero at around 1.33 μm, and is 3.2 and 17 psec/nm/km at 1.294 and 1.55 μm, respectively.[7] Figure 6.3 shows the total dispersion of the fiber.[7]

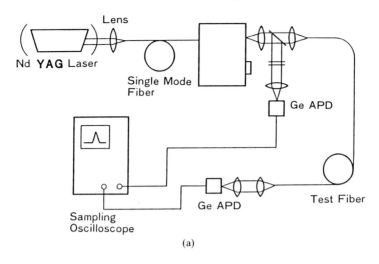

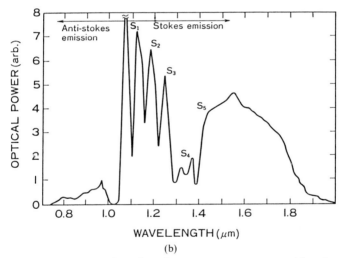

Fig. 6.2. Schematic diagram of fiber dispersion measurement apparatus, (a), and spectrum of fiber Raman laser for dispersion measurements, (b). [After Miya et al.[6.6]]

C) Dispersion-free single mode fibers in 1.5-μm region

In the conventionally designed single-mode fibers, there remains a dispersion of about 20 psec/km/nm in the 1.5-μm wavelength region, where the loss is minimum. There are two methods of overcoming this problem by modifying the fiber parameters and structure: One is to shift the zero-dispersion wavelength by changing the refractive-index difference and core

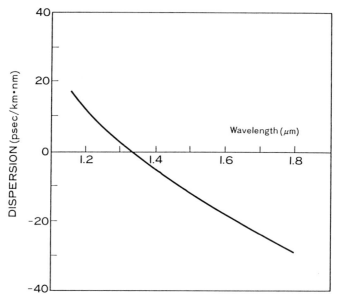

Fig. 6.3. Dispersion of a single-mode fiber, the dispersion-free wavelength; 1.3 μm. [After Kawana et al.[6,7]]

size, and the other is to use different fiber structures such as doubly clad fibers, triangular core fibers and quadruple clad fiber. In this paragraph, a dispersion-free fiber in 1.5-μm region, using the first method, will be described. A doubly clad fiber, one of the second method, will be described in the next paragraph.

The waveguide dispersion, σ, based on theoretical criteria, can be shifted towards longer wavelengths by increasing the refractive-index. Dispersion in a single-mode fiber is given by,[9]

$$\sigma = -\frac{\lambda}{C}\left[\frac{1}{2}\left\{b + (1+v)\frac{db}{dv}\right\}\frac{d^2n}{d\lambda^2} + \left(1 - \frac{1}{2}\left\{b + (1+v)\frac{db}{dv}\right\}\right)\frac{d^2n}{d\lambda^2}\right] \tag{6.1}$$

where

$a =$ core radius
$b = \dfrac{\beta^2/k^2 - n^2}{n_1^2 - n_2^2}$; normalized propagation constant

$$v = \frac{2\pi a}{\lambda}\sqrt{n_1^2 - n_2^2}\ ;\ \text{normalized frequency}$$

n_1 = refractive index in the core
n_2 = refractive index in the cladding

$$\Delta = \frac{n_1^2 - n_2^2}{2n_1^2}\ ;\ \text{relative refractive index difference}$$

β = propagation constant
k = wavenumer
c = light velocity in vacuum .

The first term is the material dispersion and the second term is the waveguide dispersion. Material dispersion has very small dependence on the dopant element and quantity. Waveguide dispersion, on the other hand, depends strongly on the waveguide structure. Therefore, it is possible to control the total dispersion through the refractive-index difference or core size.

An example of total dispersion of the fiber, with zero-dispersion wavelength at 1.55 μm, is shown in Fig. 6.4. The broken line, the dotted line and the solid line in the figure indicate material, waveguide and total dispersion, respectively. Figure 6.5 shows dispersion characteristics of three different fibers.[9] The refractive-index difference and core size of these fibers

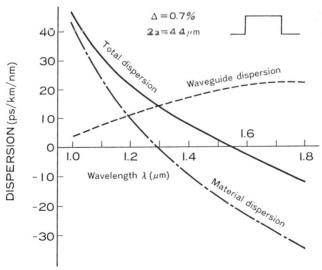

Fig. 6.4. Dispersion of a small core fiber, the dispersion-free wavelength; 1.5 μm. [After Miya et al.[6.6]]

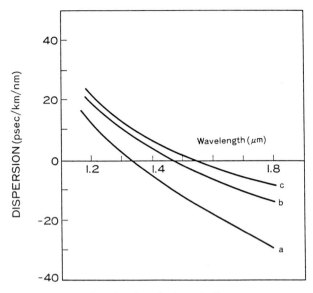

Fig. 6.5. Dispersion of various single-mode fibers.

	Refractive-index difference	Core diameter	Cut-off wavelength
a	0.20%	10.2 μm	1.04 μm
b	0.74%	5.1 μm	1.03 μm
c	0.74%	4.8 μm	1.05 μm

[After Miya et al.[6.44]]

were 0.20% and 10.2 μm for fiber a, 0.74% and 5.1 μm for fiber b, and 0.74% and 4.8 μm for fiber c.

It must be mentioned here that transmission loss of these fibers is apt to increase because the GeO_2-content is higher than that in the typical single-mode fibers.

D) Doubly clad single-mode fibers

The optical fiber structure for single-mode fibers, which exhibit low dispersion characteristics over a wide spectral range, has been proposed by Okamoto et al.[10] The fiber has two clad layers, the refractive-index of inner cladding is smaller than that of outer cladding layer, as is shown in Fig. 6.6. The typical fiber parameters are the index difference between core and outer cladding of 1% and between inner and outer cladding of −1%, the core diameter of 7.2 μm and the inner cladding thickness of 1.08 μm. Total dispersion of the fiber can be reduced to within +1 psec/km/nm over an extended range of 1.35–1.65 μm.

An interferogram and refractive-index profile of a doubly clad fiber fabricated by MCVD method is shown in Fig. 6.7.[11] The index difference,

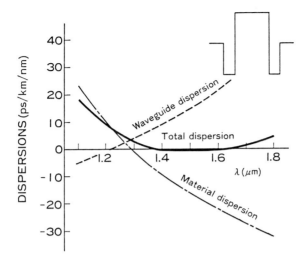

Fig. 6.6. Dispersion of doubly cladded fiber. [After K. Okamoto, Dispersionminimisation in single-mode fibers over a wide spectral range, *Electron. Lett.*, Vol. 15, No. 22, pp. 729 (1979)]

Δn_2, between the inner and outer cladding was -0.44% and the index difference, Δn_1, between the outer cladding and core was 0.61%. The ratio of the inner cladding thickness, t, to the radius of inner cladding, a, was 0.51. The preform was drawn into a fiber having a core diameter of 6.5 μm. The delay curves of the fibers measured relative to those at 1.13 μm are shown in Fig. 6.8.[11] The dispersion curve derived from the delay curve is also shown in the figure. It can be seen that the absolute value of total dispersion of this fiber is less than 1 psec/km/nm within the range of 1.32–1.43 μm.

E) *High bit rate transmission experiments*

Long-span transmission experiments of more than 100 km at 1.5 μm have been reported with extremely low-loss optical fibers, single longitudinal-mode laser diode and highly sensitive Ge-APD. Digital signal[12] at a bit rate of 445.8 Mbit/sec was transmitted through 170 km of single-mode fiber with a bit error rate less than 10^{-11}.

The fiber used in the experiments was a 170.34 km long single-mode fiber, the core diameter and index difference were 0.8 μm and 0.3%, respectively. The total loss was 43.4 dB including 17 point splices. The average fiber loss without splicing was 0.236 dB/km at 1.52 μm, and the average splice loss was 0.16 dB/splice. Chromatic dispersion was about 16 psec/km/nm.

The laser diode was injection-locked laser at a wavelength of 1.523 μm, using a distributed feedback laser diode as the master oscillator. The spectral width was about 1.5 GHz, and the main/side-mode power ratio was 35 dB.

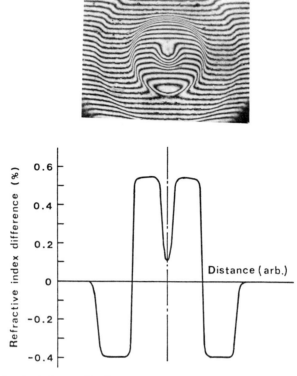

Fig. 6.7. Refractive-index profile of a doubly cladded fiber. [After T. Miya, Fabrication of low-dispersion single mode fibers over a wide spectral range, *IEEE J. Quantum Electron.*, Vol. QE-17, No. 6, pp. 858 (1981) Copyright © 1981 IEEE]

The average power launched into the fiber was 1.5 dBm, and the on/off ratio was about 14 dB.

An error rate of less than 10^{-11} was reported through the 170.34 km single-mode fiber at a received power level of -41.9 dBm. Error-rate curve for 170.34 km and 2 m of fiber and a received eye pattern after 170.34 km transmission are shown in Fig. 6.9.[12]

6.1.3 Bending loss

The transmission loss of single-mode fibers has been reduced almost to the level of the estimated loss limit. Therefore, any additional loss with cabling and installation could be significant for practical applications. Major loss increase is due to bending. Several authors have theoretically derived bending loss for single-mode fibers with a perfectly step refractive-index distribution

Transmission Characteristics of Optical Fibers 163

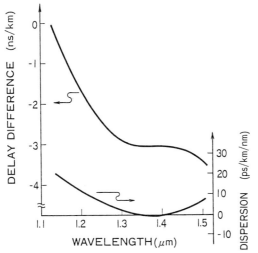

Fig. 6.8. Dispersion of a doubly cladded fiber shown in Fig. 6.7. [After Okamoto et al.[(6.10)]]

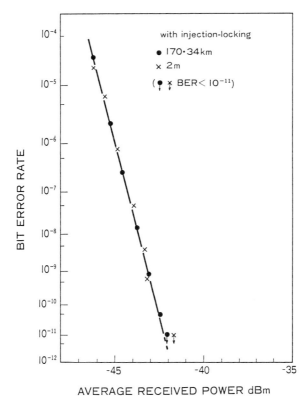

Fig. 6.9. Bit-error-rate characteristics after 170.34 km. [After H. Toba, Uniform bending losses of single-mode fibers, *Trans. IECE*, Vol. E64, No. 6, p. 331 (1980)]

and of single-mode fibers with an arbitrary refractive-index distribution using different methods. In this section, experimental results will be described.

A) Dip effect

It is very difficult to make a perfectly step refractive-index profile by the MCVD method, because of the evaporation of GeO_2 at the center of core during the preform collapse stage. Evaporation causes a dip in the refractive-index at the center of core. This dip causes an increase in bending loss.

The dip effect on bending losses is indicated in Fig. 6.10.[13] The abscissa in the figure shows the normalized dip width, that is, dip radius divided by core radius. Practical single-mode fibers fabricated by the MCVD method have a normalized dip width of less than 0.2, whose effect on bending loss is less than

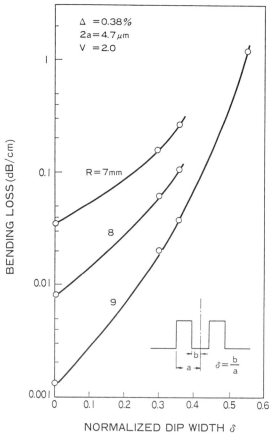

Fig. 6.10. Bending loss due to center dip of refractive-index profile. [After Kawana et al.[6.13]]

0.005 dB/km at 9 mm bending radius. When the normalized dip width is less than 0.05, the dip effect can be neglected.

B) *Bending radius and refractive-index difference*

Uniform bending losses in single-mode fibers depend strongly both on the bending radius and the refractive-index difference. Figure 6.11 shows the relation between uniform bending loss and bending radius, the parameter, Δn, in the figure is refractive-index difference.[13] Uniform bending losses decrease is proportional to the exponential function of the bending radius and also the decrease is proportional to the refractive-index difference as shown in Fig. 6.12.[13]

C) *Theory*

Several authors have derived uniform bending loss by different methods. Formulae by A. W. Snyder and D. Marcuse agree well with experimental results. The formulae can be rewritten as follows:

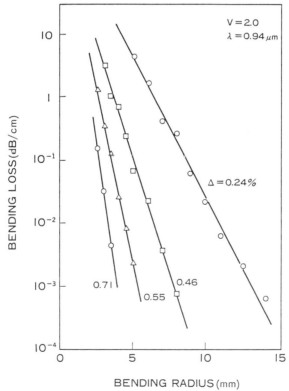

Fig. 6.11. Bending loss of single mode fibers with various index differences. [After Kawana *et al.*[6.13]]

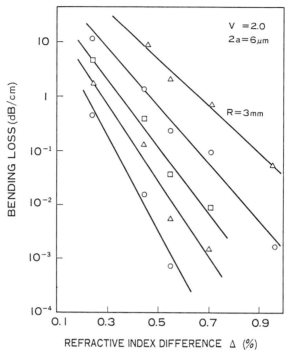

Fig. 6.12. Bending loss of single mode fibers with various bending radii. [After Kawana et al.[6.13]]

$$\alpha_1 = \frac{\sqrt{\pi}\ U^2 F}{2V^2 W\sqrt{WRa}} \cdot \frac{2W}{\pi} \qquad (6.2a)$$

$$\alpha_2 = \frac{\sqrt{\pi}\ U^2 F}{2V^2 W\sqrt{WRa}} \cdot \frac{2W}{K_1(W)} \qquad (6.2b)$$

here

$$F = \exp\left(-\frac{4\Delta W^3}{3V^2 a} R\right)$$

$$V^2 = W^2 + U^2$$

where W is the normalized decay parameter in the cladding and K_1 is the first order modified Bessel function. W can be expressed as a function of normalized frequency over the range of interest.

$$W = 1.142V - 0.9960, \qquad \text{for } 1.5 < V < 2.5. \tag{6.3}$$

Theoretical values calculated using eqs. (6.2a) and (6.2b) and experimental results of bending loss for two turns of circular bending are shown in Fig. 6.13. In Fig. 6.13, dot-dash-lines (α_2) and dash lines (α_1) indicate theoretical values, and solid lines indicate experimental results. It can be seen in this figure that α_2 shows good agreement with experimental results in the range of $2.4 < V < 1.6$ and also both α_1 and α_2 agree well with experimental results.

6.2 Graded Index High-Silica Fiber

6.2.1 Transmission loss

There are no differences in the transmission losses of high-silica graded-index fibers fabricated through OVD, MCVD and VAD processes. Figure 6.14 shows a loss spectrum for OH-free VAD fiber. Loss values at 0.85, 1.25, 1.3 and 1.6 μm wavelengths were 2.12, 0.51, 0.42 and 0.31 dB/km, respectively.[14] The low-loss window, with less than 0.5 dB/km attenuation range, was 1.2–1.7 μm wavelength. From the loss spectrum, it can be estimated that the OH-ion concentration in this fiber was about 0.8 ppb. Transmission loss of OVD fibers has also been reduced below 0.4 dB/km at 1.3 μm.[15]

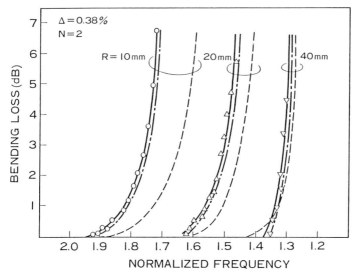

Fig. 6.13. Comparison with measured values and theoretical values for bending loss of single mode fibers. [After Kawana et al.[6.13]]

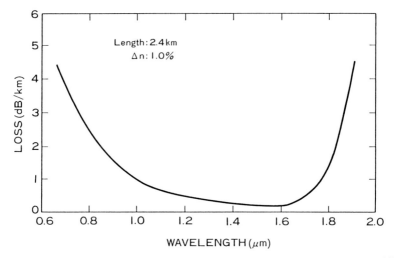

Fig. 6.14. Loss spectrum for OH-free VAD graded index fiber. [After Hanawa et al.[6,14]]

6.2.2 Bandwidth

Transmission bandwidth of graded-index fibers has been improved and the median bandwidths of OVD fibers exceed 1 GHz·km. Figure 6.15 shows a 6-dB bandwidth spectrum for a VAD fiber measured with laser diodes and fiber Raman laser. The best data so far observed is 9.3 GHz·km.[16]

6.2.3 Refractive-index fluctuation

The influence of the refractive-index dip and fluctuations on the frequency characteristics of graded-index fibers have been analyzed by several authors.[17-25] MCVD fibers and OVD fibers have sinusoidal fluctuations of refractive-index in the radial direction, and are homogeneous in the axial direction. On the other hand, VAD fibers also have sinusoidal fluctuations, furthermore, the phase of the sinusoidal index fluctuations in the radial direction varies periodically along the fiber axis. Figure 6.16 shows the interferogram of an MCVD fiber showing the index fluctuations in the radial direction. These sinusoidal index fluctuations cause transmission bandwidth degradation.

Marcuse has shown[21] that sinusoidal fluctuations superimposed on the ideal power-law profile reduce the signal bandwidth of the fiber dramatically and that phase reversal of the sinusoidal fluctuation at the midpoint of the fiber leads to substantial compensation of multimode time delay dispersion. This analysis is applicable to MCVD fibers and OVD fibers.

Okamoto has shown[25] that the bandwidth degradation due to sinusoidal fluctuations in the radial direction can be alleviated by a phase variation in the

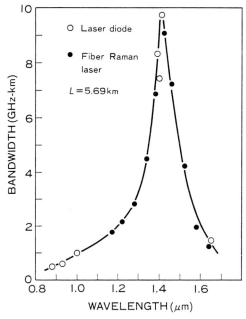

Fig. 6.15. Bandwidth spectrum for a VAD graded index fiber. [After Horiguchi et al.[6.16]]

axial directions, which is inherent to VAD fibers. Figure 6.17 shows interferograms of a VAD fiber preform, which correspond to refractive-index fluctuations in the axial direction (a) and in the radial direction (b).[26] Fluctuations were estimated to be 0.05–0.1% of the index difference between core and cladding. Refractive-index fluctuations in the radial direction due to the convex striations can be expressed as

$$n^2(r) = n_1^2[1 - 2\Delta f(r)] \qquad (6.4)$$

where

$$f(r) = (r/a)^\alpha - \varepsilon(1 - r/a)\cos(2\pi N r^2/a^2) \;;$$

α denotes the profile parameter, and ε and N are the amplitude and number of the index fluctuations, respectively. Figure 6.18 shows the bandwidth degradation due to radial fluctuation as a function of the number of fluctuations. Taking axial index fluctuation into account, the bandwidth deterioration is remarkably alleviated by the periodic phase variation along the direction of propagation, as is shown in Fig. 6.18. Actually, ultra-wide bandwidth fibers have been obtained as is shown in Section 6.2.2.

170   Chapter 6

Fig. 6.16.   Refractive-index fluctuation of MCVD fiber in radial direction.

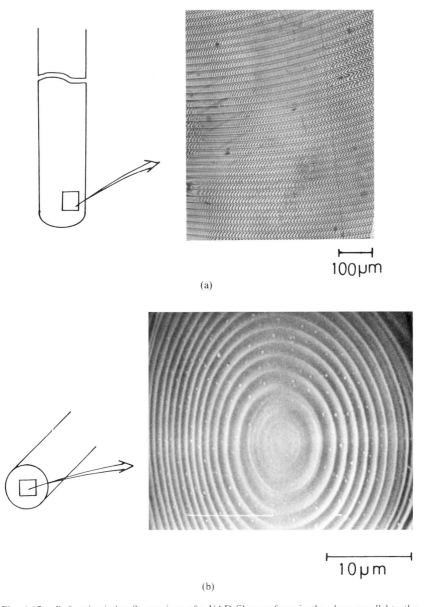

Fig. 6.17. Refractive-index fluctuations of a VAD fiber preform in the plane parallel to the growing axis, (a), and the fluctuation in the plane perpendicular to the growing axis, (b), the fluctuation is about 5×10^{-4}. [After Shibata[2.6]]

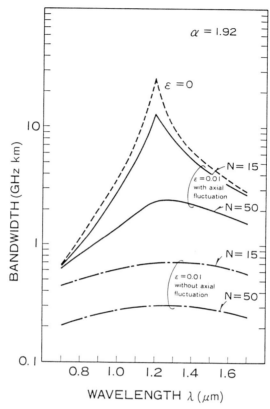

Fig. 6.18. Bandwidth of fiber with only radial fluctuation (dot-broken line) and with both radial and axial fluctuations (solid line). Broken line shows the bandwidth without fluctuations. [After K. Okamoto, Transuriesion characteristics of VAD multi mode optical fibers, *Appl. Opt.*, Vol. 20, No. 13, pp. 2314–2318 (1981)]

6.3 Al_2O_3-Doped High-Silica Fibers

Germanium oxide is the most widely used dopant to raise the refractive-index of silica glass. However, the cost of germanium is relatively high, and many efforts to use other dopant have been made. Promising candidates as the alternative dopant to germanium oxide include materials such as Al_2O_3, Sb_2O_3, ZrO_2, SnO_2 and so on. Among them, Al_2O_3 is considered most promising material, because aluminum oxide has comparatively high ability to increase the refractive-index of silica. Furthermore, fairly high purity aluminum compounds are available as the raw materials.

The fabrication of Al_2O_3-doped fibers was first reported in 1974.[27] In this work, the loss was 10 dB/km. The major loss origin was OH ions and Fe ions.

Recently, low loss Al_2O_3-doped fibers have been made by the VAD method[28] and MCVD method.[29] The source of Al_2O_3 was $AlCl_3$. The vapor pressure of $AlCl_3$ is comparatively low and it was sublimated at the temperature of 140°C, therefore, it is not easy to remove transition metal ions perfectly. Figure 6.19 shows the loss spectrum of a single mode fiber and the estimated loss factors.[30] The minimum loss was 0.65 dB/km at 1.56 μm. One half of this loss value is estimated to be due to Fe ions absorption. In order to make low-loss fibers it is important to further reduce transition metals content in the raw materials.

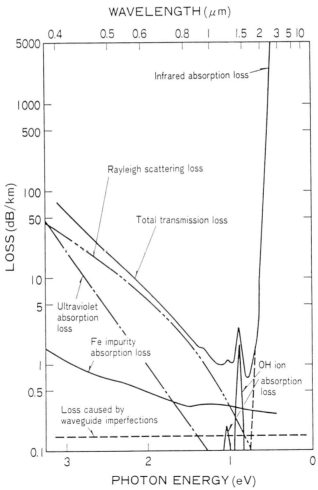

Fig. 6.19. Loss spectrum for Al_2O_3-doped silica single mode fiber. The refractive-index difference of 0.3%. [After Ohmori et al.[(6.30)]]

6.4 Loss Increase by Hydrogen Permeation

Reversible and irreversible loss increase of high-silica fibers has been observed at above 1 μm.[31,32] The origin is attributed to hydrogen permeation into the core glass. It is considered that the hydrogen molecules come from the plastic materials used for primary coating and cabling or jacketing silica glass. Figure 6.20 shows a loss spectrum of hydrogen permeated fiber. Sharp absorption peaks at 1.24, 1.17, 1.13 and 1.08 μm are attributed to the higher vibrational mode of hydrogen molecules.[31] These absorption peaks could be

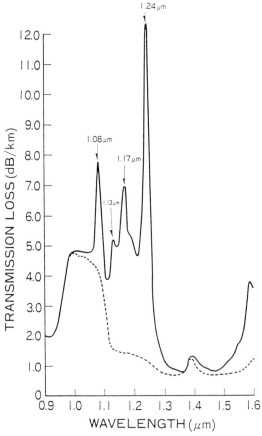

Fig. 6.20. Loss spectrum measured before (broken line) and after (solid line) placed under 1.8 atmospheres of hydrogen. [After K. Mochizuki, Transmission loss increase in optical fibers due to hydrogen permeation, *Electron. Lett.*, Vol. 19, No. 18, p. 743 (1983)]

reversibly removed by removing the hydrogen in the core glass. However, some hydrogen atoms combine with oxygen dangling bonds in the glass, to form OH ions which have strong absorption at around 1.4 μm. These OH ions can not be removed once they are formed in solid silica glass.

OH ions formed by the permeated hydrogen are strongly observed when the glass includes P_2O_5 as a constituent. Figure 6.21 shows the spectral change of phosphorus-doped MCVD fiber treated at 200°C. When the phosphorus concentration is very low, the loss increase is small as is shown in Fig. 6.22. The termination of oxygen dangling bonds with fluorine atoms reduces the formation of OH ions with permeated hydrogen.

6.5 Multi-Component Glass Fibers

Intensive attempts have been made to reduce the transmission loss of multi-component glass fibers. Figure 6.23 shows a loss spectrum of a SiO_2–GeO_2–Na_2O–CaO–Li_2O–MgO glass fiber.[33] The major loss origin is the presence of transition metal ions as shown in the figure. In this case the impurities of iron and chromium ions make a large contribution to the transmission loss increase. It is estimated that these transition metal ions come into the glass through the raw material alkali metal oxides and alkaline-earth metal oxides. The minimum loss so far reported is below 4 dB/km at 0.8–0.9

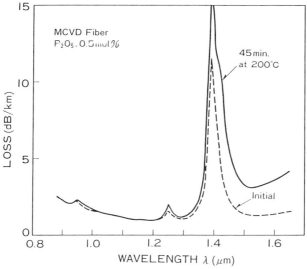

Fig. 6.21. Example of loss spectra for a nylon jacketed fiber fabricated by MCVD method. [After N. Uesugi, Optical loss increase of phosphor-doped silica fiber at high temperature in the long wavelength region, *Appl. Phys. Lett.*, Vol. 43, No. 4, p. 432 (1984)]

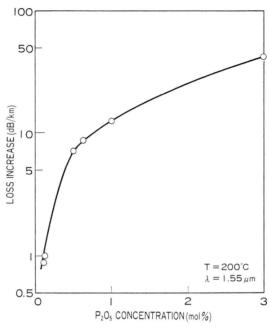

Fig. 6.22. Saturated values of loss increase at 1.55 μm as a function of P_2O_5 concentration. Heating temperature is 200°C. [After Uesugi et al.[6.32]]

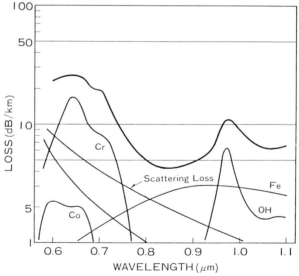

Fig. 6.23. Transmission loss of a multi-component glass fiber and major loss origins. [After S. Takahashi, Low-loss and loss-OH content soda-lime-silica glass fiber, *Electron. Lett.*, Vol.4, No. 5, p. 151 (1978)]

μm wavelength region.[33] It is not so easy to reduce transition metal ions any further. The reduction of hydroxyl ions is also very important to fabricate low-loss fibers.

6.6 Non-Silica Fibers

6.6.1 Fluoride glass fibers

Fluoride glass fibers are highly transparent in the 1.5–2.5 μm wavelength region, where the transmission loss of high-silica fibers increase. As is the case of multi-component glass fibers, the major loss in fluoride glass fibers is due to transition metal ions and hydroxyl ions. The fundamental OH vibration introduces a strong absorption at 2.9 μm, the OH ions reduction is an important approach to realize low-loss fibers.

Figure 6.24 shows a loss spectrum of a BaF_2–GdF_3–ZrF_4–AlF_3 glass fiber fabricated by a "built-in casting" process. The minimum loss is 6.3 dB/km at 2.13 μm.[34]

The chemical durability of fluoride glasses is fairly poor compared to oxide glass such as flint glass. Figure 6.25 shows the rate of weight reduction of a fluoride glass and multi-component oxide glasses in water at various temperature.[35] Therefore, strict protection from water vapor is necessary for practical applications of fluoride fibers.

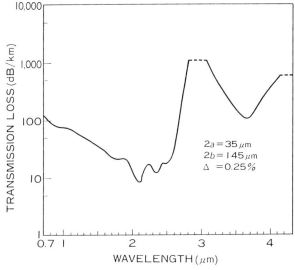

Fig. 6.24. Loss spectra of fluoride glass fiber. [After S. Mitachi, Fabrication of low OH and low loss fluoride optical fiber, *Jpn. J. Appl. Phys.*, Vol. 23, No. 9, p. L726 (1984)]

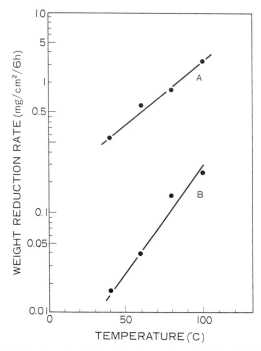

Fig. 6.25. Weight reduction rate of a fluoride glass and a oxide glass. A: 31.68BaF$_2$-3.84GdF$_3$-60.48ZrF$_4$-4.0AlF$_3$, B: 45.7SiO$_2$-3.6Na$_2$O-5.0K$_2$O-45.1PbO-0.6As$_2$O$_3$. [After Mitachi et al.[6.35]]

6.6.2 Chalcogenide glass fibers

Chalcogenide glass fibers are usable at the wavelength region from 1 μm to 8 μm, where the transmission loss is below 1000 dB/km. The lowest loss so far reported is 35 dB/km at 2.44 μm for an As$_{40}$S$_{60}$ unclad fiber.[36] Figure 6.26 shows the loss spectrum. The peak at 2.91 μm arises from the OH fundamental stretching vibration. The strong peak at 4.03 μm is due to the SH fundamental stretching vibration.[37] The strong absorptions in other chalcogenide glass fibers are also due to SH and SeH vibrations. Figures 6.27(a), (b) show the transmission loss of these fibers.[36] In these materials, a weak absorption tail is clearly observed, and it is the major cause of the loss limit at shorter wavelength region. The peaks observed are mostly attributed to SH, SeH, OH stretching vibrations and their combinations. The loss limit of these fibers, when the SH, SeH and OH impurity ions are totaly removed, is estimated from the spectra. The estimated loss limit of As$_{40}$S$_{60}$, As$_{38}$Ge$_5$Se$_{57}$ and Ge$_{20}$S$_{80}$ fibers are 23 dB/km at 4.6 μm, 30 dB/km at 6.3 μm and 169 dB/km at 3.6 μm.[38]

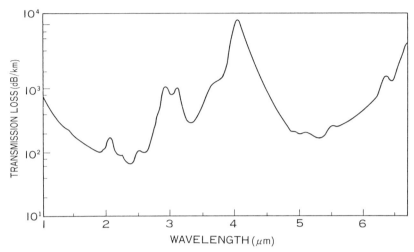

Fig. 6.26. Transmission loss spectrum for an $As_{40}S_{60}$ unclad fiber with 200 μm diameter. [After Kanamori et al.[1,34]]

6.6.3 Crystalline fibers

Crystalline fibers are estimated to have very low loss, however, the actual fibers made of various crystalline materials have comparatively high transmission loss. They still have the advantage of a very low loss in the infrared region longer than 5 μm.

The most widely studied crystalline fibers are extruded polycrystalline fibers made of KRS-5 (thallium bromoidite). The transmission loss at 10.6 μm[39] is less than 100 dB/km and it increases at the shorter wavelength as λ^{-2}. Though single-crystal fibers have a high potential as ultralow-loss fibers, it is very difficult to fabricate them with low loss. The reason is not only the presence of impurities but also surface irregularities and other structural imperfections. The losses for CsI and CsBr single crystal fibers with a plastic loose cladding are 80,000 dB/km[40] and 5000 dB/km,[50] respectively.

6.6.4 Plastic fibers

Figures 6.28(a) and (b) show the transmission spectra of polymethylmethacrylate (PMMA) fiber and polystyrene fiber, respectively.[42] The cladding material was fluoroalkyl methacrylate polymer, whose refractive-index is lower than that of PMMA and polystyrene. The polymer has sufficient adhesion ability to the core materials. The absorption peaks denoted as v_n^0, v_n^1 and δ in the figures are attributed to the stretching vibrations of aliphatic hydrocarbons, the stretching vibrations of aromatic hydrocarbon and bending vibrations, respectively. The subscript n shows the n-th higher mode

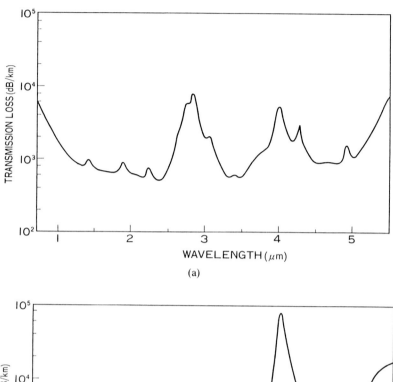

Fig. 6.27. Transmission loss spectra for a $Ge_{20}S_{80}$ unclad fiber, (a), and $As_{38}Ge_5Se_{57}$ unclad fiber, (b), both with 200 μm diameter. [After Kanamori et al.[1.34]]

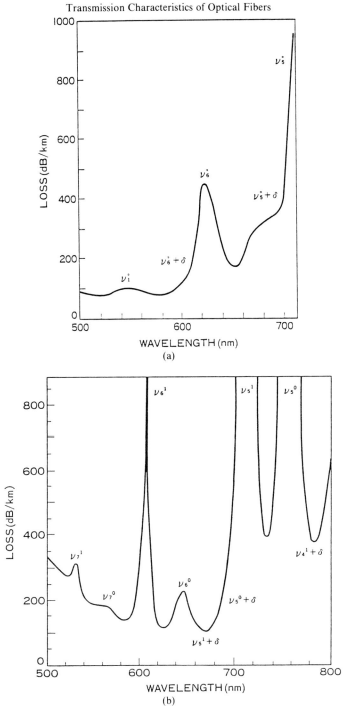

Fig. 6.28. Transmission loss spectra for PMMA core plastic fiber, (a), and polystylene core plastic fiber, (b). [After Kaino et al.[1.34]]

of the fundamental vibration. Rayleigh scattering is another major loss factor. The scattering loss values of PMMA and polystyrene are 13 dB/km and 55 dB/km at 6328 Å, respectively. The experimental value of PMMA coincides to the calculated value from the density fluctuation using Clausius-Mossotti's equation. The minimum loss of PMMA fiber was 55 dB/km at 0.58 μm, as is shown in Fig. 6.28 and it is estimated that the loss is composed of 7 dB/km absorption loss, 20 dB/km of Rayleigh scattering loss and 28 dB/km of structural imperfection loss.

Hydrocarbon vibrational absorption in polymers can be reduced by substituting carbon atoms to deuterium atoms without changing the chemical characteristics. By substituting hydrogen with deuterium, fundamental frequency of the CD molecular vibration shifts to 1.35 times longer wavelength region than CH.[43] The 6th harmonic of CD vibration appears at 840 nm with an intensity of 120 dB/km, whereas the 6th harmonic of CH vibration appears at 622 nm with an intensity of 440 dB/km. Thus, the deuteration not only causes the vibrational absorption to shift to a longer wavelength, but also a decrease in absorption intensity. Figure 6.29 shows the loss spectrum of a deuterated PMMA core fiber.[42] The residual hydrogen in the deuterated PMMA was less than 0.7%. The lowest loss of 20 dB/km was attained from 650 to 670 nm. One half of the loss is attributed to the structural imperfections such as core diameter fluctuations and core-cladding interface imperfections. Therefore, the loss limit of PMMA fiber is estimated to be 10 dB/km.

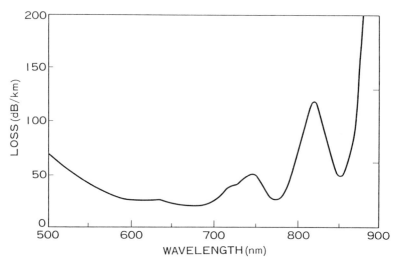

Fig. 6.29. Transmission loss spectrum for poly-deuterated-MMA core plastic fiber. [After Kaino et al.[1,34]]

REFERENCES

1) T. Miya, "Study on low-loss and low-dispersion single mode fibers" (in Japanese), Ph. D. Thesis, Tohoku Univ. (1983).
2) T. Miya, Y. Terunuma, T. Hosaka and T. Miyashita, "Ultimate low-loss single-mode fiber at 1.55 μm", Electron. Lett., Vol. 15, pp. 106-108, Feb. (1979).
3) R. Yamauchi, K. Kosaka, M. Miyamoto and O. Fukuda, "Low-loss VAD single mode fibers for 1.55 μm region", National Convention Rec. IECE Jpn., p. 443 (Mar. 1984).
4) K. Kosaka, Y. Igarashi, T. Moriyama and O. Fukuda, "Los-loss low-OH-content single-mode fiber", Nat. Convention Rec. IECE Japan, Paper 1144 (1984).
5) R. Csencsits, P. J. Lemaire, W. A. Reed, D. S. Shenk and K. L. Walker, "Fabrication of low-loss single-mode fibers", Tech. Dig. OFC'84, pp. 54-55 (1984).
6) T. Miya, M. Nakahara and N. Inagaki, "VAD single-mode fiber fabrication techniques— 1.5 μm dispersion-free single-mode fibers", Rev. ECL, Vol. 32, No. 3, pp. 411-417 (1984).
7) A. Kawana, T. Miya, Y. Terunuma, and T. Hosaka, "Fabrication of dispersion-free single-mode fibers around 1.5 μm wavelength region", Trans. IECE, Vol. E64, No. 9, p. 591 (1981).
8) D. Gloge, "Dispersion in weakly guiding fibers", Appl. Opt., Vol. 10, No. 11, p. 2442 (1971).
9) A. Sugimoto, K. Daikoku, N. Imoto and T. Miya, "Wavelength dispersion characteristics of single-mode fibers in low loss region", IEEE J. Quantum Electron., Vol. QE-16, No. 2, p. 215 (1980).
10) K. Okamoto, T. Edahir, A. Kawana and T. Miya, "Dispersioniminmization in single-mode fibers over a wide spectral range", Electron. Lett., Vol. 15, No. 22, p. 729 (1979).
11) T. Miya, K. Okamoto, Y. Ohmori and Y. Sasaki, "Fabrication of low-dispersion single mode fibers over a wide spectral range", IEEE J. Quantum Electron., Vol. QE-17, No. 6, p. 858 (1981).
12) H. Toba, Y. Kobayashi, K. Yanagimoto, H. Nagai and M. Nakahara, "Injection-locking technique applied to a 170 km transmission experiment at 445.8 Mbit/s", Electron. Lett., Vol. 20, No. 9, p. 370 (1984).
13) A. Kawana, T. Hosaka and T. Miya, "Uniform bending losses of single-mode fibers", Trans. IECE, Vol. E64, No. 6, p. 331 (1980).
14) F. Hanawa, S. Sudo, M. Kawachi and M. Nakahara, "Fabrication of completely OH-free VAD fiber", Electron. Lett., Vol. 16, No. 18, pp. 669-700 (1981).
15) C. W. Deneka, "Update on advances in the Outside Vapor Deposition process", Tech. Dig. Conf. Opt. Fiber Comm. (OFC'84), New Orleans (1984).
16) M. Horiguchi, M. Nakahara, N. Inagaki, K. Kokura and K. Yoshida, "Transmission characteristis of ultra-wide bandwidth VAD fibers", Tech. Dig. 8th ECOC, AIII-4, p. 75 (1982).
17) J. A. Arnaud and W. Mammel, "Dispersion in optical fibers with starlike refractive-index profiles", Electron. Lett., Vol. 12, No. 1, p. 6 (1976).
18) R. Olshansky, "Pulse broadening caused by deviations from the optical index profile", Appl. Opt., Vol. 15, No. 3, p. 782 (1976).
19) D. Marcuse, "Calculation of bandwidth from index profiles of optical fibers I: Therory", Appl. Opt., Vol. 18, No. 12, p. 2073 (1979).
20) D. Marcuse, "Multimode fiber with z-dependence value", Appl. Opt., Vol. 18, No. 13, p. 2229 (1979).
21) D. Marcuse, "Multimode delay compensation in fibers with profile distortions", Appl. Opt., Vol. 18, No. 23, p. 4003 (1979).
22) T. Matsumoto, "Effects of index-profile distortion and splicing peremutation on fiber bandwidth" (in Japanese), Paper of Tech. Group, IECE Jpn., Vol. TG-OQE 79-134, pp. 31-36 (1980).

23) Y. Daido, "Refractive-index profiling of graded-index fiber waveguides attaining 100 MHz bandwidth at 100 km", CLEOS, 26–28 Feb. 1980, Dig. Tech. Papers (Opt. Soc. Am., Washington, D.C., 1980), paper TUDD 3.
24) C. Pask, "Pulse propergation in optical fibers with index profiles slowly varying along their length", *Opt. Quantum Electron.*, Vol. 12, p. 281 (1980).
25) K. Okamoto, T. Edahiro and M. Nakahara, "Transmission characteristics of VAD multimode optical fibers", *Appl. Opt.*, Vol. 20, No. 13, pp. 2314–2318 (1981).
26) N. Shibata, "Optical properties of high-silica glass for optical fibers" (in Japanese), Ph. D. Thesis, Nagoya Univ. (1982).
27) S. Kobayashi, H. Nakagome, N. Shimizu, H. Tsuchiya and T. Izawa, "Low-loss optical glass fiber with Al_2O_3–SiO_2 core", *Electron. Lett.*, Vol. 10, No. 20, p. 410 (1974).
28) Y. Ohmori, F. Hanawa and M. Nakahara, "Fabrication of low-loss Al_2O_3-doped silica fibers", *Electron. lett.*, Vol. 18, p. 761 (1982).
29) J. R. Simpson and J. B. MacChesney, "Optical fibers with an Al_2O_3-doped silicate core composition", *Electron. Lett.*, Vol. 19, No. 7, p. 261 (1983).
30) Y. Ohmori, M. Nakahara and M. Horiguchi, "Properties of alumina doped VAD fibers", *Rev. ECL*, Vol. 32, No. 3, p. 432 (1984).
31) K. Mochizuki and Y. Namihira, "Transmission loss increase in optical fibers due to hydrogen permeation", *Electron. Lett.*, Vol. 19, No. 18, p. 743 (1983).
32) N. Uesugi, T. Kuwabara, Y. Koyamada, Y. Ishida and N. Uchida, "Optical loss increase of phosphor-doped silica fiber at high temperature in the long wavelength region", *Appl. Phys. Lett.*, Vol. 43, No. 4, p. 327 (1983).
33) S. Takahashi, S. Shibata and M. Yasu, "Low-loss and low-OH content soda-lime-silica glass fiber", *Electron. Lett.*, Vol. 4, No. 5, p. 151 (1978).
34) S. Mitachi, Y. Ohishi and S. Takahashi, "Fabrication of low OH and low loss fluoride optical fiber", *Jpn. J. Appl. Phys.*, Vol. 23, No. 9, p. L726 (1984).
35) N. Mitachi, Y. Ohishi, Y. Terunuma and S. Takahashi, "Fabrication of fluoride glass fibers" (in Japanese), *ECL Tech. J.*, Vol. 32, No. 12, p. 2732 (1983).
36) T. Kanamori, Y. Terunuma, S. Takahashi and T. Miyashita, "Chalcogenide glass fibers for mid-infrared transmission", *IEEE J. Light Tech.*, Vol. LT-2, No. 5, p. 607 (1984).
37) P. A. Young, "Optical properties of vitrous arsenic trisulfide", *J. Phys. C: Solid St. Phys.*, Vol. 4, p. 93 (1971).
38) T. Kanamori, Y. Terunuma and S. Takahashi, "Preperation of chalcogenide optical fiber", *Rev. ECL*, Vol. 32, No. 3, p. 469 (1984).
39) S. Kachi, M. Kimura, H. Kikuchi and K. Shiroyama, "Effect of surrounding gasses in polycrystal infrared fiber" (in Japanese), Tech. Dig. Nat. Conf. Jpn. Soc. Appl. Phys., 12p-E-16, Oct. (1984).
40) T. J. Bridges, J. S. Hasiak and A. R. Strand, "Single-crystal AgBr infrared opticcal fibers", *Opt. Lett.*, Vol. 5, No. 3, p. 85 (1980).
41) Y. Mimura, Y. Okamura, Y. Komazawa and C. Ota, "Growth of fiber crystals for infrared optical waveguides", *Jpn. J. Appl. Phys.*, Vol. 19, No. 5, p. L269 (1980).
42) T. Kaino, M. Fujiki and K. Jinguji, "Preperation of plastic optical fibers", *Rev. ECL*, Vol. 32, No. 3, p. 478 (1984).
43) H. M. Schleinitz, "Ductile plastic optical fibers with improved visible and near infrared transmission", *Int. Wire & Cable Symp.*, Vol. 26, p. 352 (1977).
44) T. Miya, A. Kawana, T. Terunuma and T. Hosaka, "Fabrication of single-mode fibers for 1.5 μm wavelength region", *Trans. IECE Jpn.*, Vol. E63, p. 514 (1980).
45) K. Okamoto and T. Miya, "Zero total dispersion in single-mode optical fibers over an extended spectral region", *Radio Sci.*, Vol. 17, p. 31 (1982).
46) N. Shibata, "Optical characteristics of high silica glass fibers for communication", Ph. D. Thesis, Nagoya Univ. (1983).

INDEX

alkali metal oxide, 5
As_2S_3, 16
$As_{38}Ge_5Se_{57}$, 178
$As_{40}S_{60}$, 178

B_2O_3, 27
B_2O_3–GeO_2 doped glass, 28
B_2O_3–SiO_2, 27
bandwidth, 168
BCl_3, 56
bending loss, 162
Brillouin scattering, 9, 19
brominating reagent, 112
bubble formation, 141

$C_2Cl_2F_2$, 112
calcogenide glass, 6
carbon resistance furnace, 67
chalcogenide glass, 18, 148, 178
chelate resin, 138
chlorinating reagent, 112
cladding, 90
cladding-to-core diameter ratio, 92
combination tone, 25
consolidation furnace, 82
critical pore diameter, 103
crystalline fiber, 179
CsI, 18
CVD, 54

dehydration, 94, 107
deposition rate, 94
deuterated PMMA, 13, 151
dip, 164
doubly clad, 160
drawing apparatus, 67

EFG, 150
electron transition absorption, 15
extrusion, 150

fine glass particle, 87
fluctuation, 168
fluoride glass, 6
fluoride glass fiber, 177
fluorinating reagent, 112
fluorine doping, 131

gas permeability, 105
$Ge_{20}S_{80}$, 178
$Ge_{33}As_{12}Se_{55}$, 16
$GeCl_4$, 56
GeO_2, 26
GeO_2-doped silica, 30, 35
glass-working lathe, 56

halogenating process, 108
hydrogen permeation, 174
hydroxyl impurity, 61
hydroxyl ion, 32

ISD method, 54

KCl, 18
KRS-5, 150, 179

Landau-Placzek ratio, 19

material dispersion, 38, 41, 159
material dispersion variant, 46
MCVD, 52, 54
minimum loss wavelength, 36
modal dispersion, 38
Modified Chemical Vapor Deposition, 52
multiphonon, 11
multiphonon absorption, 13

network former, 5
network modifier, 5

OH contamination, 62

Outside Vapor Deposition process, 52, 64
OVD, 52
overtone, 25
oxide glass, 4
oxy-hydrogen flame, 86

P_2O_5, 26
P_2O_5–SiO_2, 27
phonon absorption, 11, 12
plastic coating applicator, 67
PMMA, 8, 151, 179
$POCl_3$, 56
polycrystal, 8
polycrystalline fiber, 150
porous preform, 79, 85, 88
prebake, 58
primary coating, 71
profile dispersion, 41
profile parameter, 44, 127
pulling machine, 83

Raman scattering, 9, 20
Rayleigh scattering, 9, 19, 33
reaction chamber, 81
refractive-index, 36
Reynold's number, 94

scattering loss, 18

Sellmeier's equation, 38
Sellmeier's parameter, 44
$SiCl_4$, 56
$SiHCl_3$, 61, 86
single crystal, 6
single-mode fiber, 92
sintering, 97
SiO_2–B_2O_3, 119
SiO_2–GeO_2, 119, 122
SiO_2–TiO_2, 119, 122
$SOBr_2$, 112
$SOCl_2$, 82, 108, 112
surface treatment, 69

thermal expansion coefficient, 46
torch, 80
total dispersion, 159
transition metal oxide, 5
transparent preform, 79
trichlorosilane, 61

Urbach's rule, 15
UV-absorption tail, 30

VAD, 52, 77
Vapor-phase Axial Deposition, 52

waveguide dispersion, 159
weak absorption tail, 16